Environmental Review Guidance for Licensing Actions Associated with NMSS Programs

Final Report

U.S. Nuclear Regulatory Commission
Office of Nuclear Material Safety and Safeguards
Washington, DC 20555-0001

Environmental Review Guidance for Licensing Actions Associated with NMSS Programs

Final Report

Manuscript Completed: July 2003
Date Published: August 2003

Division of Waste Management
Office of Nuclear Material Safety and Safeguards
U.S. Nuclear Regulatory Commission
Washington, DC 20555-0001

ABSTRACT

This guidance document provides general procedures for the environmental review of licensing actions regulated by the Office of Nuclear Material Safety and Safeguards (NMSS). Although the main focus of this guidance is the NRC staff's environmental review process, it also contains related information which applicants and licensees may find useful. Chapter 1 provides a summary and overview of the guidance. This chapter briefly discusses the three ways in which an environmental review is performed: either by meeting the criteria for a categorical exclusion or by preparing an environmental assessment (EA) or environmental impact statement (EIS). This chapter also discusses early planning for an EA or EIS and methods of using previous environmental analyses related to the proposed action. Chapter 2 discusses the categorical exclusions and the basis of their use. Chapter 3 discusses the EA process, including preparation and content of the EA, and preparation of the Finding of No Significant Impact. Chapter 4 discusses the process of preparing an EIS, from developing a project plan through scoping, consultations and public meetings, to preparing the Record of Decision. Chapter 5 discusses the technical content of the EIS, and Chapter 6 discusses environmental information that should be considered by applicants and licensees in preparing their environmental report.

Paperwork Reduction Act

Public Protection Notification

iv

CONTENTS

Appendices

Figures

Tables

ACRONYMS/ABBREVIATIONS

ADAMS	Agencywide Documents Access and Management System
ALARA	As Low As is Reasonably Achievable
BLM	Bureau of Land Management
CATX	Categorical Exclusion
CEQ	Council on Environmental Quality
CFR	Code of Federal Regulations
DEIS	Draft Environmental Impact Statement
DOE	U.S. Department of Energy
DWM	Division of Waste Management (NRC)
EA	Environmental Assessment
EIS	Environmental Impact Statement
EPA	U.S. Environmental Protection Agency
EPAB	Environmental and Performance Assessment Branch
ER	Environmental Report
FEIS	Final Environmental Impact Statement
FONSI	Finding of No Significant Impact
FR	*Federal Register*
FWS	U.S. Fish and Wildlife Service
GEIS	Generic Environmental Impact Statement
NEPA	National Environmental Policy Act
NMSS	Office of Nuclear Material Safety and Safeguards
NRC	U.S. Nuclear Regulatory Commission
OGC	Office of General Counsel
OFA	Office of Federal Activities
OSTP	Office of State and Tribal Programs
PM	Project Manager
RAI	Request for Additional Information
RCRA	Resource Conservation and Recovery Act
ROD	Record of Decision
SER	Safety Evaluation Report
SHPO	State Historic Preservation Officer
TAR	Technical Assistance Request
THPO	Tribal Historic Preservation Officer
WWW	World Wide Web

CITATIONS FOR LAWS AND REGULATIONS

This document uses accepted abbreviations for referencing the United States Code, the Code of Federal Regulations, and the *Federal Register*.

United States Code (USC)

The format for United States Code is xx USC yyyy, where xx represents the title and yyyy represents the section. For example, the Atomic Energy Act can be found at 42 USC 2011, et seq. The Latin phrase, *et seq. (et sequentes)* literally means "and the following." *Et seq.* can be interpreted to mean "and the subsequent sections." The United States Code is available on the WWW at <http://www.access.gpo.gov/congress/cong013.html>.

Code of Federal Regulations (CFR)

The format for the Code of Federal Regulation is xx CFR yyy, where xx represents the title and yyy represents the part. For example, the U.S. Nuclear Regulatory Commission regulations on environmental protection can be found at 10 CFR 51. The Code of Federal Regulations is available on the WWW at <http://www.access.gpo.gov/nara/index.html>.

Federal Register (FR)

The format for the *Federal Register* is xx FR yyyy, where xx is the volume number and yyyy is the page number. For example, the U.S. Nuclear Regulatory Commission's final rule for license termination criteria is found at 62 FR 39058. The *Federal Register* is available on the WWW at <http://www.access.gpo.gov/nara/index.html>.

1 INTRODUCTION TO THIS GUIDANCE DOCUMENT

1.1   Introduction

The National Environmental Policy Act (NEPA) of 1969 (42 USC 4321 et seq.) requires Federal agencies, as part of their decision-making process, to consider the environmental impacts of actions under their jurisdiction. Both the Council on Environmental Quality (CEQ) and the U.S. Nuclear Regulatory Commission (NRC) have promulgated regulations to implement NEPA requirements. CEQ regulations are contained in the Code of Federal Regulations (CFR) at 40 CFR 1500 to 1508, and NRC requirements are provided in 10 CFR 51.[1]

To ensure consistent treatment of environmental requirements throughout the NRC Office of Nuclear Material Safety and Safeguards (NMSS), the Environmental and Performance Assessment Branch (EPAB) has produced this document to provide general procedures for determining the level of environmental review and documentation required for NMSS actions. **Because of broad NMSS organizational responsibilities (e.g., rulemaking, licensing of new facilities, amendments to existing licenses, and decommissioning), this document is written in general terms to accommodate the different situations and types of facilities regulated by NMSS**. Divisions within NMSS and their Regional counterparts may have, or choose to develop, supplemental environmental review guidance that is specific to facilities they regulate. Any such guidance should be consistent with this document and provided to EPAB for review and comment. This document is intended to provide an overview of the environmental review process with major emphasis on preparing NEPA documents. Chapter 2 discusses categorical exclusions (CATXs), Chapter 3 discusses the environmental assessment (EA) process, Chapters 4 and 5 discuss the environmental impact statement (EIS) process and technical content, respectively, and Chapter 6 discusses environmental information that should be considered by applicants and licensees in an environmental report (ER).

This document is primarily intended to serve as guidance to NMSS staff to meet the requirements established by legislation and regulations. Although this guidance is not a substitute for legislation and regulations and compliance with this document is not required, the NMSS staff is directed to use this guidance when reviewing licensing actions. In a similar manner, applicants and licensees are encouraged, but not required, to use Chapter 6 when preparing environmental reports for submission to the NRC. Methods different from those set out in this document will be acceptable if they provide a basis for concluding that the NRC's regulations have been met.

This document supersedes previous environmental review guidance, including:

*   NMSS Policy and Procedures Letter 1-48, Revision 1, "Procedure for Preparing Environmental Assessments," May 31, 1995.

*   NMSS Policy and Procedures Letter 1-50, Revision 2, " Environmental Justice in NEPA Documents," September 7, 1999.

---

[1]While the NRC maintains its view that, as a matter of law, independent regulatory agencies can be bound by the CEQ NEPA regulations only insofar as those regulations are procedural or ministerial in nature, the regulations nonetheless provide useful guidance. See 49 FR 9352.

- Memorandum from Richard E. Cunningham, "Policy and Guidance Directive 84-20; Impact of Revision of 10 CFR Part 51 on Material Licensing Actions," December 5, 1984.

- Memorandum from Richard E. Cunningham, "Supplement to Policy and Guidance Directive 84-20: Impact of Revision of 10 CFR Part 51 on Material Licensing Actions," February 19, 1992.

- Memorandum from Carl J. Paperiello, "Revision 1, Supplement to Policy and Guidance Directive 84-20 'Impact of Revision of 10 CFR Part 51 on Material Licensing Actions'," March 9, 1994.

- Memorandum from John T. Greeves, "Guidance on Preparation of Environmental Assessments for Licensing Actions by Regional Offices," May 7, 2001.

1.2   The Environmental Review

NEPA mandates that Federal agencies carefully consider the environmental impacts of their actions prior to making decisions that affect the environment. The NEPA review (also referred to as environmental review) process is usually initiated by an application for a new license or certification, change to an existing license, or a decommissioning plan submitted to the NRC. A flow chart illustrating the NEPA process is shown in Figure 1. Part of the NEPA process is directed by legislation and Executive Orders related to environmental issues. When a request for a specific action is received from an applicant/licensee, the NRC first determines whether a CATX is applicable for the proposed action. CATXs are categories of actions that the NRC, in consultation with CEQ, has determined do not individually or cumulatively have a significant effect on the environment. Criteria for identifying a CATX and a list of actions eligible for CATX are provided in 10 CFR 51.22. Categories of actions appropriate for CATX include administrative, organizational, or procedural amendments to certain types of NRC regulations, licenses, and certificates; minor changes related to application filing procedures; certain personnel and procurement activities; and activities where environmental review by NRC is excluded by statute. If a CATX exists, the finding should be documented as described in Chapter 2. The proposed action is subject to no further NEPA review, but is still evaluated for compliance with NRC radiation protection regulations and other applicable environmental regulations. Under special circumstances, the NRC may elect to conduct an environmental review even if a CATX exists [10 CFR 51.22(b)].

If no CATX applies, NMSS staff responsible for the proposed action must prepare an EA (10 CFR 51.21). EAs are prepared by the appropriate licensing project manager[2] (PM) or rulemaking task leader and are reviewed by an environmental PM in EPAB. An EA is a concise publicly available document that provides sufficient evidence and analysis for determining whether to prepare an EIS or a finding of no significant impact (FONSI). If the EA supports a FONSI, the environmental review process is complete. If the EA reveals the proposed action may significantly affect the environment and cannot be mitigated, the environmental review activities transition to the process to develop an EIS.

---

[2]In this document the term "licensing PM" is used to refer to the staff responsible for preparing the EA. Different organizations within NMSS or the Regions may utilize staff with different titles to assume responsibility for completion of the EA.

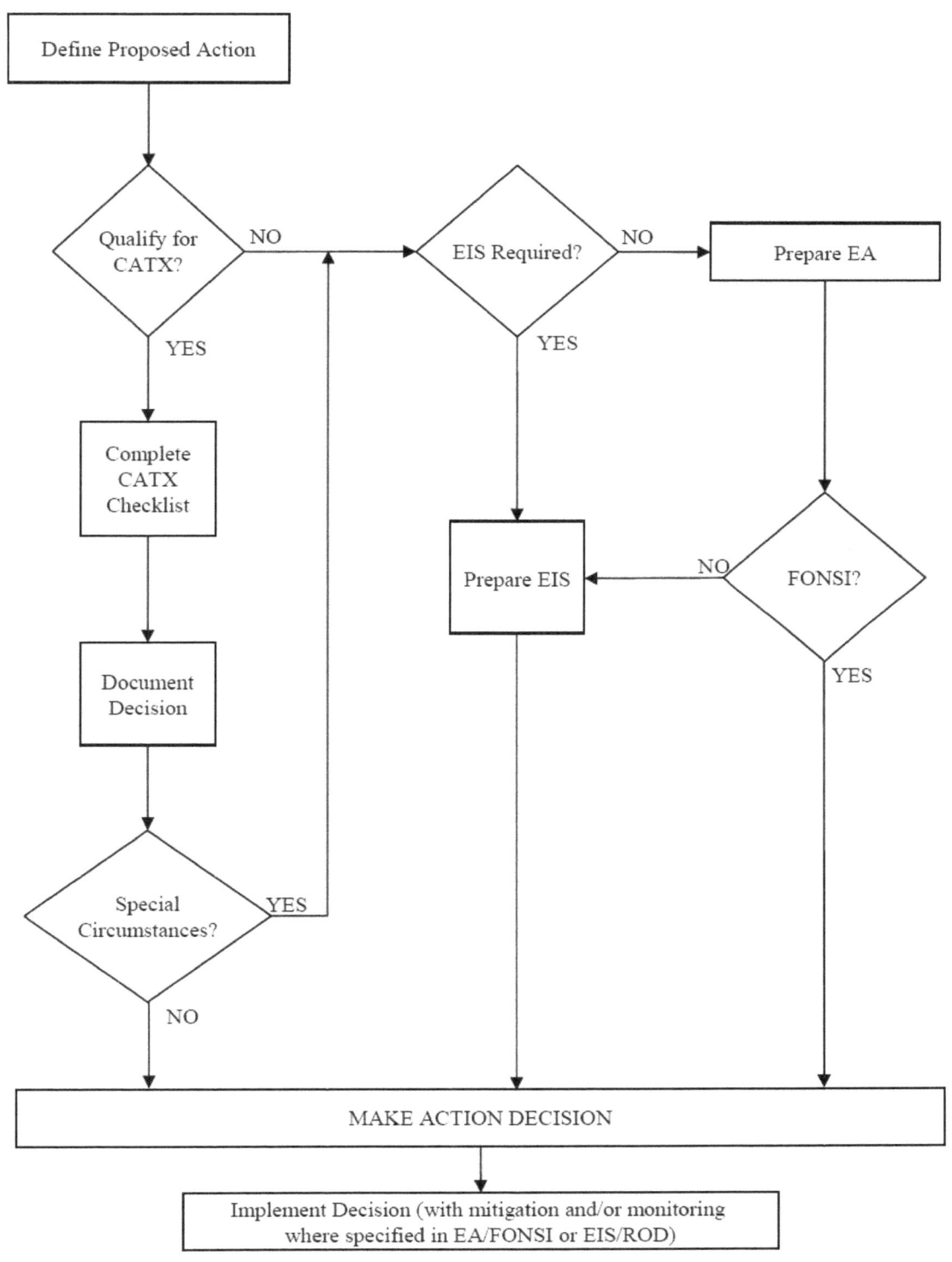

Figure 1: Flow chart showing NEPA screening process.

NEPA requires that a detailed statement of the environmental impact of the proposed action and reasonable alternatives be prepared for "major Federal actions significantly affecting the quality of the human environment" [Section 102(2)(C)]. NRC implementing regulations require an EIS for proposed actions that are major Federal actions significantly affecting the quality of the human environment or involve a matter which the NRC determines should be covered by an EIS [10 CFR 51.20(b)]. An EIS is also prepared for actions in which an EA does not support a FONSI. An EIS is a publicly available document detailing the environmental impacts associated with the proposed action and reasonable alternatives. Except for some rulemaking EISs, all NMSS EISs are prepared by environmental PMs in EPAB (EISs prepared in support of rulemaking should be reviewed by EPAB). For NMSS, types of licensing actions that typically require an EIS are applications for facilities such as uranium mills, uranium conversion plants, uranium enrichment plants, independent spent fuel storage installations at a site not occupied by a nuclear power reactor, and low-level waste disposal facilities.

The Rulemaking and Guidance Branch in the Division of Industrial and Medical Nuclear Safety (IMNS) is responsible for preparing EAs and EISs for rulemaking actions in NMSS. Procedures for EAs and EISs prepared in support of rulemakings are described in NUREG/BR-0053, *Regulations Handbook* (NRC, 2001a). Related guidance documents, (i) NMSS Policy and Procedures Letter 1-63, *Procedures for Preparation and Review of Rulemaking Packages* (NRC, 2001b) and (ii) Management Directive 6.3, *The Rulemaking Process* (NRC, 2000), should be used in conjunction with this guidance document. EPAB management should be informed of any proposed rulemaking actions that will require an EIS early in the rulemaking and planning stage. EPAB is required to review all EAs and EISs prepared for rulemaking actions (NOTE: EA's dealing with certificates of compliance do not need to be reviewed by EPAB if the previously reviewed template is followed).

The focus of environmental review documents (e.g., EAs and EISs) is the environmental impacts of the proposed action and reasonable alternatives. Often, the NRC staff prepares a Safety Evaluation Report (SER) to evaluate and document the safety of the proposed action and compliance with NRC regulations. The safety and environmental reviews are conducted in parallel. Although there is some overlap between the content of an SER and the NEPA document, the intent of the documents are different. The NEPA document does not address accident scenarios, rather it addresses the environmental impacts which would result from the accident and is therefore dependent on certain information from the SER. Accident scenarios (i.e., frequency, probability) are addressed in the SER. Much of the information describing the affected environment is also applicable to the SER (e.g., traffic patterns, demographics, geology, and meteorology) and the NRC staff should ensure consistency between the NEPA document and the SER. For rulemakings, SERs are not developed and there is generally no site-specific information nor environmental report.

1.3 Initiation of the Environmental Review

An environmental review may be initiated as a result of a license application, request for a licensing action, or as a result of an internal action, such as a rulemaking. Any of these activities may result in the documentation of a CATX or the preparation of an EA or an EIS, depending on the significance of impacts associated with the action. The environmental review needs to be coordinated, scheduled, and integrated with the safety evaluation and other aspects of the licensing or certification action.

Preparation of an EA for a licensing action for which a CATX does not apply is the responsibility of the licensing PM either at NRC Headquarters or in a Regional Office. All EAs prepared for NMSS actions must be reviewed by EPAB[3]. EPAB staff can also assist with the various consultations among other Federal and State Agencies, as discussed in Sections 1.4, 3.3, and 3.4.9.

If the EA results in a FONSI, it is also the licensing PM's responsibility to prepare the *Federal Register* notice of the EA finding and the basis for those findings. If the EA does not result in a FONSI, the EIS process should be initiated as discussed in Chapter 4. Please note that all *Federal Register* notices related to materials licensing actions must be reviewed by OGC (NRC, 2002a). Additionally, guidance is currently being developed to assist staff in the preparation of *Federal Register* notices.

1.3.1 Early Planning

If the proposed licensing action is likely to require preparation of an EA, the licensing PM should consult with EPAB to determine the level of EPAB assistance required. For complex licensing actions, as described in Section 3.4, the licensing PM should involve EPAB as early as possible to facilitate the EA review. A written request should be forwarded to EPAB requesting review of the EA. Chapter 6 provides guidance to assist the applicant/licensee in preparing an environmental report (ER) which aids the NRC in preparing the EA or EIS and complying with Section 102(2) of NEPA. The general requirements for a materials applicant/licensee ER are provided in 10 CFR 51.60.

Written requests for EPAB reviews of EAs should include the following information:

- A generalized description of the proposed action(s);

- Copy of the SER, license application, or background material, as applicable;

- Requested date for completion of EPAB's review; and

- The activity code to be entered for Time and Labor in the NRC Human Resource Management System (HRMS) for the requested work.

All requests for EA reviews or other requests for review assistance (e.g., review the use of a CATX or assist in any required consultation) be sent to the electronic mail address Environmental_Reviews@nrc.gov (please note there is an underscore between Environmental and Reviews). EPAB management will notify the requestor and identify the environmental PM working on the request and the estimated time of completion. The typical review period for EAs is 30 days. To request an early or accelerated review, please include a rationale and propose a completion review date. EPAB will review the request and respond whether the request is possible.

The responsible licensing PM should continue to keep EPAB informed about significant changes and issues, new information, and meetings that are planned regarding the proposed action. The licensing PM may also request assistance from EPAB in preparing correspondence for any required consultations. The licensing PM should keep in mind that consultations with other Federal, State, local, and tribal agencies

---

[3]In the future, after Headquarters and Regional staff gain more experience preparing EAs consistent with this guidance, only "complex" EAs will need to sent to EPAB. However, if requested, EPAB will review "simple" EAs. EPAB will issue a memorandum notifying the staff of changes in the review requirements for EAs.

may be required and should be initiated early in the environmental review as discussed in Sections 1.4, 3.3, and 3.4.9. In addition, EPAB may need copies of information related to the EA (e.g., request for the proposed action, information provided by the applicant/licensee, and referenced documents used in tiering). In this case, EPAB staff will work with the licensing PM to identify and obtain copies of the necessary information.

If it is determined that an EIS is necessary, a Technical Assistance Request (TAR) should be forwarded requesting EPAB to assume responsibility for the EIS. The TAR should include the same information as suggested above in the EA. The environmental PM in EPAB will establish a separate activity code for Time and Labor in HRMS for preparation of an EIS. The environmental PM will need copies of all information related to the EIS (e.g., applicant/licensee request for the proposed action, supporting information provided by the applicant/licensee, referenced documents used in tiering, and previous site specific EAs and SERs applicable to the proposed action). At this point, the environmental PM will prepare a notice of intent for the *Federal Register* to inform the public of the decision to prepare an EIS (10 CFR 51.26-27). Section 4.2.2, *Notice of Intent*, provides more information on this notice.

Consultations with other Federal, State, local, and tribal agencies may be required and should be identified by the licensing PM and initiated early during the environmental review. These consultations can take a significant amount of time and may identify impacts which require further assessment.

1.3.2 Pre-Application Meetings

Prior to the applicant/licensee's preparation of information needed to support the environmental review (e.g., information usually found in an ER), the licensing PM should be aware that preliminary meetings between the applicant/licensee and EPAB can enhance the environmental review process (10 CFR 51.40).

1.3.3 Acceptance Reviews

An application or request for action is accompanied by information needed to conduct the environmental analysis. This information may be provided in an ER submitted by the applicant/licensee. Information may also be submitted as part of the license application or amendment request, without an ER. When the environmental information is submitted, the licensing PM should conduct an acceptance review of the applicant/licensee information to determine whether (i) the requested action qualifies for a CATX or whether an EA or EIS is required, and (ii) the information is complete and will support the required environmental analyses. This initial acceptance review is not be a detailed technical review and does not address quality and thoroughness of the ER; rather, the acceptance review determines if the submitted information is complete in order to begin the detailed technical review. For actions that might require an EIS, the licensing PM should involve EPAB in the acceptance review. The information in Chapter 6, *The Environmental Report: Format and Technical Content* provides a list of topics that may be helpful in completing the preliminary assessment of the submittal.

The appropriate PM (i.e., licensing or environmental depending on whether an EA or EIS is being prepared) should ensure that the applicant/licensee has provided data, assumptions, and analyses, that support the applicant/licensee conclusions. The appropriate PM should begin to develop an outline of the environmental review document (EA/EIS) during the acceptance review in order to reveal gaps in explanations or logic that may require additional information from the applicant/licensee. If the environmental information is deficient, the applicant/licensee should be notified by letter that deficiencies in the submittal prevent NRC from beginning the review.

Once the submittal is determined to be acceptable for the environmental review, the applicant/licensee should be notified by letter that the NRC has begun its review. The letter should also notify the applicant/licensee that, in the course of the detailed review, the staff may identify areas where additional information is needed to complete the review. The letter should also provide the applicant/licensee with a time frame for the completion of the staff's review.

1.4   Consultations

When receiving a license application, the PM should be aware of the following consultations and the time required to complete these consultations. It should be noted that even if the proposed action is categorically excluded from NEPA review, the PM should determine if the proposed action requires consultation under either Section 7 of the Endangered Species Act of 1973 (16 USC 1531 et seq.) or Section 106 of the National Historic Preservation Act of 1966 (16 USC 470 et seq.). If so, the PM should proceed with the appropriate consultation as described below. In addition to the overview presented below, a step-by-step procedure is provided in Appendix D.

PMs are encouraged to coordinate compliance with Section 7 and Section 106 with any steps taken to meet the requirements of NEPA. The PM should consider their consultation responsibilities as early as possible in the NEPA process, and plan any required public participation, analysis, or review in such a way that they can meet the purposes and requirements of all three statutes in a timely and efficient manner.

Consistent with applicable conflict of interest laws the PM may obtain information from the applicant/licensee or use the services of consultants to prepare information, analyses or recommendations in completing these consultations. However, the NRC remains responsible for the reliability of information used and all required findings and determinations. If a document or study is prepared by a non-Federal party, the PM is responsible for ensuring that its content meets applicable standards and guidelines.

For rulemaking actions, consultations will not typically be required as these types of actions are usually considered administrative in nature. However, the rulemaking task leader should consult the procedures provided in Appendix D.

The PM should also be aware of NRC's commitment to the Council on Environmental Quality to consult with affected States before issuing the environmental assessment (NRC, 1994). Additional information is provided in Section 3.3, 3.4.9, and Appendix D.

1.4.1 Section 106 of the National Historic Preservation Act

Section 106 of the National Historic Preservation Act requires the NRC staff to take into account the effects of the undertaking (i.e., the licensing action) on historic properties, and afford the Advisory Council on Historic Preservation a reasonable opportunity to comment. If the proposed action meets the criteria for an "undertaking" or has the potential to cause effects to historic properties, consultation with the State Historic Preservation Officer (SHPO) is required (36 CFR 800). The review should also consider historic properties included in State or local registers or inventories and any additional important cultural, traditional, or historic properties. In areas of American Indian tribal land, a Tribal Historic Preservation Officer (THPO) may exist with the equivalent authority and responsibility as a SHPO. Responses from the SHPO/THPO should be noted in the EA or included in an appendix to the EIS.

1.4.2 Section 7 of the Endangered Species Act

Section 7 of the Endangered Species Act (Interagency Cooperation) requires the NRC staff to ensure that the licensing action is "not likely to jeopardize the continued existence of any endangered species or threatened species or result in the destruction or adverse modification of the habitat of such species." If a proposed action "may affect" listed species or critical habitat, consultation with the U.S. Fish and Wildlife Service (FWS) or National Marine Fisheries Service is required (50 CFR 402). Additional information can be found in "Endangered Species Consultation Handbook" (FWS, 1998). Responses from either agency should be noted in the EA or included in an appendix to the EIS.

1.5 Emergency Circumstances

Under certain circumstances, the NRC may approve a licensee's action which could result in significant environmental impacts without first observing the provisions of 10 CFR 51. This situation will only occur under emergency circumstances where the health and safety of the public may be adversely affected if mitigative or remedial actions are delayed (10 CFR 51.13). In these cases, the EPAB will consult with the CEQ as soon as practicable concerning alternative NEPA arrangements.

1.6 Utilizing Existing Environmental Analyses

Existing environmental analyses should be considered to evaluate impacts associated with a proposed action to the extent possible and appropriate. This approach builds on work that has already been done, avoids redundancy, and provides a coherent and logical record of the analytical and decision making process.

The staff should determine whether environmental analyses (EAs or EISs) relevant to the site or proposed action have been prepared and whether any of the existing analyses adequately address the proposed action and alternatives. This review will determine whether additional analyses are necessary or whether a tiered or supplemental analysis is possible.

For example, an initial license application or license renewal EA may be very detailed and provide many different aspects of anticipated activities for that license. If subsequent license actions are identified specifically or are bound by the initial analysis in the original license application or renewal EA, the environmental review for subsequent licensing actions may incorporate by reference the initial analysis and additional NEPA evaluation may be limited to consideration of new and significant information. When utilizing a previous environmental review the licensing PM should ensure this is documented either by a note to the appropriate docket or license file or in a letter to the applicant of licensee. The following language should be considered in instances where no additional environmental review is necessary:

"The NRC staff have previously reviewed the environmental impacts from this license amendment in **[insert title of document, date, and description of scope]**. The scope of this license amendment was included in the previous analysis and does not alter the previous **[FONSI or EIS decision, insert date published in *Federal Register*]**. No further environmental review is required under 10 CFR Part 51.

1.6.1 Adopting

Adoption is another technique used to avoid redundancies in NEPA analysis. If the NRC elects to use all or part of another agency's EIS, the NRC can adopt the EIS as described in the CEQ regulations (40 CFR 1506.3, 10 CFR 51 Appendix A). For example, under the Nuclear Waste Policy Act, the NRC is required, to the extent practicable, to adopt the U.S. Department of Energy's (DOE) EIS for a proposed high-level radioactive waste geologic repository (10 CFR 51.26).

In those instances where the actions covered by another agency's EIS and the NRC action are substantially the same, the NRC can adopt the EIS after recirculating the document as a FEIS. Section 4.9, *Publishing the FEIS*, provides more information on publishing the FEIS. When recirculating the FEIS, the NRC should provide information that identifies the Federal action involved. The EIS must meet the applicable NRC criteria and the NRC must prepare a separate record of decision (ROD).

In those instances where the actions covered by the other agency's EIS and the NRC proposal are not substantially the same, the NRC can adopt the EIS in part by treating the other agency's final document as part of an NRC DEIS (discussed in Section 4.4, *Publishing the DEIS*). If the other agency's EIS only partially covers a proposed action or only a portion of the other agency's EIS is adopted, the NRC must prepare a supplemental or new and separate DEIS, describing that portion of the other agency's EIS which is being adopted as well as any supplementary analysis needed. If the NRC adopts an EIS that is not final within the agency that prepared it, or if the adequacy of the EIS is the subject of judicial action that is not final, the NRC must indicate its status in the recirculated DEIS and/or FEIS (40 CFR 1506.3) and a FEIS and ROD must be prepared.

Similar procedures exist for using another agency's EA. The other agency's EA must satisfy CEQ and NRC criteria. The NRC takes full responsibility for the scope and content of any adopted EA. The NRC must prepare its own FONSI (in accordance with 10 CFR 51.32–35) and decision record. Another agency's FONSI and decision record cannot be used or adopted by the NRC.

An example of adoption is the NRC's use of the DOE's Generic Environmental Impact Statement (GEIS) for the management of spent nuclear fuel (DOE, 1995). This 1995 GEIS examined the various programmatic alternatives for dealing with spent nuclear fuel and also identified certain site-specific actions. Subsequent to the GEIS, the DOE applied for a site-specific 10 CFR 72 license to store the Three Mile Island, Unit 2, spent nuclear fuel that was discussed and analyzed in the 1995 GEIS. As part of the environmental review, the NRC adopted DOE's 1995 GEIS, with minor additions, and recirculated it as a DEIS and FEIS (NRC, 1998). For this licensing action NRC determined that site-specific aspects had been addressed and, therefore, it was appropriate to adopt the 1995 GEIS.

1.6.2 Tiering

Tiering (defined in 40 CFR 1508.28) is a procedure by which more specific or more narrowly focused environmental documents can be prepared without duplicating relevant parts of previously prepared, more general, or broader documents. The new, more specific environmental document incorporates by reference the general discussions and analyses from the existing broader document and concentrates on the issues and impacts of the project which are not specifically covered in the broader document. Often, the broader document is referred to as a programmatic EIS or GEIS. The new environmental document, however, must be within the scope and conclusions of the more general environmental document from which it is tiered. Also, the decision made as a result of the more specific document does not change or modify the decision(s) of the more general document. The new environmental document must identify the document from which it is tiered and both documents must be available for public review. An example of tiering may include using a GEIS as the basis for an EA or EIS prepared for a site-specific proposed action.

Since NEPA documents prepared for rulemaking are usually generic in nature, tiering off previous documents is usually inappropriate. However, future applicant/licensee proposals requiring NEPA reviews should be assessed for possible tiering from a rulemaking EA or GEIS. Therefore, an initial rulemaking NEPA document, especially a GEIS, should provide ample information regarding bounding conditions and assumptions to allow future reference and tiering. An example of tiering off a rulemaking GEIS is provided in Appendix A. In this example, a checklist was developed to assist in the determination of whether the GEIS in support of the License Termination Rule (NRC, 1997), is applicable to proposed decommissioning actions.

An example of tiering is the NRC's use of the DOE's GEIS for the management of spent nuclear fuel (DOE, 1995). This GEIS examined the various programmatic alternatives for dealing with spent nuclear fuel and also identified certain site-specific actions. Subsequent to the GEIS, the DOE, through a private contractor, applied for a site-specific 10 CFR 72 license to store Peach Bottom spent nuclear fuel. This particular action was discussed in the GEIS but did not include site-specific details. The NRC elected develop an EIS (62 FR 48953) that tiers off DOE's GEIS (i.e., incorporates relevant portions into the current environmental review while developing site-specific impact analyses). For this licensing action, NRC determined that site-specific aspects had not been adequately addressed in the GEIS and it was therefore not appropriate to adopt the GEIS. Another specific example of tiering is the use of NUREG-1437, "Generic Environmental Impact Statement for License Renewal of Nuclear Plants," (NRC, 1996a) as it applies to the spent fuel storage license renewal actions.

1.6.3 Supplementing

A supplement to an existing draft EIS (DEIS) or final EIS (FEIS) is prepared when additional environmental analysis is needed as described in 10 CFR 51.72 or 51.92.

It is not necessary to formally supplement an EA. An existing EA can be modified to reflect new information. For example, a modified EA could be prepared by identifying changes to an existing EA and attaching or incorporating by reference the existing EA.

1.6.4 Incorporating by Reference

Incorporation by reference is a technique used to avoid redundancies in analyses and to reduce the bulk of a NEPA document. Both EAs and EISs may incorporate previous analyses by reference. Materials or analyses incorporated by reference are not limited to NEPA documents. Special technical or professional studies and analyses prepared by the NRC, other Federal, State, local agencies, tribal governments, or private interests may be incorporated by reference.

The EA or EIS should identify documents that are incorporated by reference and indicate where these references are available for public review. Relevant portions of the incorporated analysis should be referenced by page or section number and summarized in the EA or EIS. Incorporating by reference should not result in a loss of comprehension to the reader. The NEPA document must be able to stand alone and provide sufficient analysis to allow the decision maker to arrive at a conclusion. Material incorporated by reference must be reasonably available for inspection by interested persons within the time allowed for comment. Material based on proprietary data may not be incorporated by reference.

An example of incorporation by reference is the NRC's use of the DOE's GEIS for the management of high-level waste (DOE, 2002). In the NRC's preparation of an EIS for a site-specific 10 CFR 72 license to store Peach Bottom spent nuclear fuel, the NRC was able to incorporate by reference the affected environment section of the 2002 DOE GEIS as the two actions are at the same location.

1.7 Public Meetings

In preparing for meetings PMs should be aware of NRC's "Enhancing Public Participation in NRC Meetings; Policy Statement" (67 FR 36920; NRC, 2002b). Additional guidance is available for conducting public meetings in NUREG/BR-0224, "Guidelines for Conducting Public Meetings" (NRC, 1996b) and NUREG/BR-0297, "NRC Public Meetings" (NRC, 2002c). Relevant guidance is also contained in NRC Management Directive 3.5 "Public Attendance at Certain Meetings Involving NRC Staff" (NRC, 1996c).

1.8 Sensitive Information

In preparing environmental review documents, the PM should be aware of certain types of information that may be restricted for national security reasons or eligible for withholding under other specific statutory provisions. The PM is referred to 10 CFR 51.16 and NRC Management Directive 3.4 "Release of Information to the Public" (NRC, 1999) for additional details.

There may also be occasions where an EA or EIS is required for a proposed action that is classified. These documents must be restricted from public dissemination for national security reasons. These documents should be organized so that classified information is included in an appendix that is not made publically available while unclassified portions can be made available to the public (40 CFR 1507.3(c)).

1.9 References

DOE (U.S. Department of Energy), 1995. "Department of Energy Programmatic Spent Nuclear Fuel Management and Idaho National Engineering Laboratory Environmental Restoration and Waste

Management Programs Final Environmental Impact Statement." DOE/EIS-0203. DOE, Washington, DC. April.

DOE, 2002. "Idaho High Level Waste and Facilities Disposition Final Environmental Impact Statement." DOE/EIS-0287. DOE, Washington, DC. September.

FWS (U.S. Fish and Wildlife Service), 1998. "Endangered Species Consultation Handbook." FWS, United States Department of Agriculture, Washington, DC. <http://endangered.fws.gov/consultations/index.html>. (December 19, 2002).

NRC (U.S. Nuclear Regulatory Commission), 1994. "State Consultation on Environmental Assessments." Memorandum from Taylor to Russell et. al. U.S. Nuclear Regulatory Commission, Washington, DC. December 6.

NRC, 1996a. "Generic Environmental Impact Statement for License Renewal of Nuclear Plants." NUREG-1437. U.S. Nuclear Regulatory Commission, Washington, DC. May.

NRC, 1996b. "Guidelines for Conducting Public Meetings." NUREG/BR-0224. U.S. Nuclear Regulatory Commission, Washington, DC. February.

NRC, 1996c. "Public Attendance at Certain Meetings Involving the NRC Staff." Management Directive 3.5. U.S. Nuclear Regulatory Commission, Washington, DC. May 24.

NRC, 1997. "Generic Environmental Impact Statement in Support of Rulemaking on Radiological Criteria for License Termination of NRC-Licensed Facilities." NUREG-1496. U.S. Nuclear Regulatory Commission, Washington, D.C.

NRC, 1998. "Final Environmental Impact Statement-For the Construction and Operation of an Independent Spent Fuel Storage Installation to Store the Three Mile Island Unit 2 Spent Fuel at the Idaho National Engineering and Environmental Laboratory." NUREG-1626. U.S. Nuclear Regulatory Commission, Washington, D.C.

NRC, 1999. "Release of Information to the Public." Management Directive 3.4. U.S. Nuclear Regulatory Commission, Washington, DC. December 1.

NRC, 2000. "The Rulemaking Process." Management Directive 6.3. U.S. Nuclear Regulatory Commission, Washington, D.C. June 2.

NRC, 2001a. "Regulations Handbook." NUREG/BR-0053, Revision 5. U.S. Nuclear Regulatory Commission, Washington, D.C. March.

NRC, 2001b. "Procedures for Preparation and Review of Rulemaking Packages." NMSS Policy and Procedures Letter 1-63. U.S. Nuclear Regulatory Commission, Washington, D.C. August.

NRC, 2002a. "*Federal Register* Notices For Materials Licensing Actions." Memorandum from Travers to Meserve. U.S. Nuclear Regulatory Commission, Washington, D.C. December 10.

NRC, 2002b. "Enhancing Public Participation in NRC Meetings; Policy Statement." U.S. Nuclear Regulatory Commission, Washington, DC. May 28. <http://www.nrc.gov/public-involve/public-meetings/meeting-faq.html>. (December 19, 2002).

NRC, 2002c. "NRC Public Meetings." NUREG/BR–0297. U.S. Nuclear Regulatory Commission, Washington, D.C. August.

PAGE INTENTIONALLY BLANK

2 PREPARATION AND USE OF CATEGORICAL EXCLUSIONS

The purpose of CATXs is to focus extensive NEPA analysis onto major Federal actions that may significantly affect the quality of the human environment. The use of CATXs is a means of streamlining the NEPA process, saving time, effort, and resources.

Categorical exclusion "...means a category of actions which do not individually or cumulatively have a significant effect on the human environment and which have been found to have no such effect in procedures adopted by a Federal agency [...] and for which, therefore, neither an Environmental Assessment nor an Environmental Impact Statement is required..." (40 CFR 1508.4).

NRC regulations further describe CATXs in 10 CFR 51.22. A list of current categorical exclusion criteria can be found at 10 CFR 51.22(c). Under special circumstances, the NRC may issue an EA or EIS for any action which is categorically excluded. In the final rule, the Commission declined to further define special circumstances because of the difficulty in precisely defining future situations and to maintain necessary flexibility (49 FR 9352). Special circumstances include, but are not limited to, unresolved conflicts concerning alternative uses of available resources within the meaning of section 102(2)(E) of NEPA. Special circumstances in which a CATX may not apply are discussed further in the following sections.

2.1   Documenting the CATX

All CATXs should be documented in some manner. This documentation provides the explicit evidence that the staff carried out the NEPA process and provides the rationale for applying the CATX. At a minimum the CATX should be documented in the safety or technical review or a letter of response to the applicant/licensee noting which CATX applies and how it applies. For actions which clearly qualify for a CATX, no coordination with EPAB is necessary and the following sentence should be included in the response to the applicant/licensee or in a memo to the file (without further documentation):

> "An environmental assessment for this action is not required, since this action is categorically excluded under 10 CFR 51.22(c)(*licensing manager should fill in appropriate number*)."

For rulemaking actions, the CATX is documented in the Statement of Considerations. More detail is provided in NUREG-0053, "Regulations Handbook" (NRC, 2001).

For actions not clearly encompassed by the CATX, additional documentation should be placed in the license file. The additional documentation could be information supplied by the applicant/licensee or a note to the license file from the licensing PM describing why or how a CATX applies. Examples of additional information to determine whether a CATX applies can be found in NUREG-1556, Vol. 20, "Consolidated Guidance About Materials Licenses: Guidance About Administrative Licensing Procedures" (NRC, 2000a). It is suggested that the licensing PM coordinate with EPAB in making determinations on whether a CATX is applicable.

This chapter also provides an acceptable method of documenting a CATX via a checklist. Appendix B contains a generic CATX checklist with instructions. This generic checklist can also be used to document whether special circumstances are present. Although additions to the checklist are allowed, the five basic questions, in the checklist, should be answered for each CATX.

Also, Sections 51.22(c)(11) and 51.22(c)(14)(xvi) provide generic categorical exclusions. The use of either CATX was addressed in a Staff Requirements Memorandum (NRC, 1984) which directed the staff to prepare:

> "a written memorandum explaining why the action qualifies for the categorical <u>exclusion</u> (emphasis in original) selected. The written memorandum shall include a discussion of the factors listed in the selected subsections[4] and shall become part of the permanent docket or record relating to that action."

This 1984 memo also directed that the explanatory memo be signed by the appropriate Division Director (or designee) and included in the license file.

2.2   General CATX Guidance

On March 12, 1984, 10 CFR 51, "Environmental Protection Regulations for Domestic Licensing and Related Regulatory Functions," was revised (49 FR 9352). Subsequent guidance highlighted the following licensing actions which are not covered by categorical exclusion:

- Use of radioactive tracers in field flood studies involving secondary and tertiary oil and gas recovery.

- Performance of field studies in which licensed material is deliberately released directly into the environment for purposes of the study. However, the use of tracers in well-logging is specifically covered by 10 CFR 51.22(c)(14)(xi).

- Processing of source material for extraction of rare earth and other metals.

- Waste brokers who are authorized to store waste more than 180 days, or possess more than 50 curies of radioactive material.

- Any commercial radioactive waste disposal.

The following sub-sections provide text for the most commonly used CATXs in NMSS and the basis for the CATX as noted in the Discussion and Finding in the final rule which created the CATX *(NOTE: Excerpts from the appropriate Statements of Consideration published with the final rule are shown in italics)*. When a licensing or rulemaking action is not clearly encompassed by the Discussion and Findings, additional documentation must be provided in the license file describing how or why the CATX applies. If it is determined that the CATX does not apply (e.g., special circumstance), an EA must be developed. EPAB and the Office of General Counsel (OGC) staff may be consulted when making the determination of whether a CATX applies and the nature and extent of the documentation to support a CATX.

---

[4]The "selected subsections" are 10 CFR 51.22(c)(9), (c)(11), or (c)(14)(xvi).

2.2.1 Background for 10 CFR 51.22(c)(1)

This CATX states:

> (1) Amendments to Parts 1, 2, 4, 7, 8, 9, 10, 11, 19, 21, 25, 55, 75, 95, 110, 140, 150, 170, or 171 of this chapter, and actions on petitions for rulemaking relating to Parts 1, 2, 4, 7, 9, 10, 11, 14, 19, 21, 25, 55, 75, 95, 110, 140, 150, 170, or 171.

Discussion and Finding (49 FR 9352)

The regulations in these parts serve the dual purpose of making needed information readily available to the public and providing procedures for the orderly conduct of Commission business. These regulations in and of themselves will not affect the volume of that business.

In some instances, the regulations implement Federal laws and executive orders which prescribe specific procedures and policies for the conduct of government business. These laws include the Administrative Procedure Act (15 U.S.C. 551 et seq.), the Freedom of Information Act (5 U.S.C. 552), the Privacy Act of 1974 (Pub. L. 93-579), the Government in the Sunshine Act (5 U.S.C. 552b), the Federal Advisory Committee Act (Pub. L. 92-463, 86 Stat. 770), certain provisions in 18 U.C.C. 201-209 dealing with conflicts of interest in Federal employment, and House Concurrent Resolution No. 175, July 11, 1958, on the Code of Ethics for Government Service (72 Stat. B12, 5 U.S.C.A. § 7301, Note.) Executive Order 11222, May 8, 1965, provides in part that "[t]he elimination of conflict of interest in the Federal service is one of the most important objectives in establishing general standards of conduct."

In some instances, application of the regulations will have economic or social, but not environmental consequences. Examples include: Part 140 which contains regulations implementing the provisions of the Price- Anderson Act relating to financial protection and indemnity agreements; Part 170 which prescribes the schedule of Commission fees; and Part 4 which contains regulations on nondiscrimination which implement the provisions of Title VI of the Civil Rights Act of 1964 and Title IV of the Energy Reorganization Act of 1974.

Formal interpretations of the Commission's regulations authorized by the Commission and prepared by the General Counsel are codified in Part 8. Although these interpretations may address matters of substance as well as procedure, the issuance of a formal interpretation and its inclusion in Part 8 of the Commission's regulations is an action without environmental effect.

2.2.2 Background for 10 CFR 51.22(c)(2)

This CATX states:

> (2) Amendments to the regulations in this chapter which are corrective or of a minor or nonpolicy nature and do not substantially modify existing regulations, and actions on petitions for rulemaking relating to these amendments.

Discussion and Finding (49 FR 9352)

Minor amendments of this type are sometimes needed to update, clarify or eliminate an ambiguity in an existing regulation. Since these amendments are usually editorial and do not change the substance of an existing regulation they can neither increase nor decrease any environmental impact which the existing regulation may have.

2.2.3 Background for 10 CFR 51.22(c)(3)

This CATX states:

> (3) Amendments to parts 20, 30, 31, 32, 33, 34, 35, 39, 40, 50, 51, 54, 60, 61, 70, 71, 72, 73, 74, 81 and 100 of this chapter which relate to—(i) Procedures for filing and reviewing applications for licenses or construction permits or other forms of permission or for amendments to or renewals of licenses or construction permits or other forms of permission; (ii) Recordkeeping requirements; or (iii) Reporting requirements; and (iv) Actions on petitions for rulemaking relating to these amendments.

Discussion and Finding (49 FR 9352)

Although amendments of this type affect substantive parts of the Commission's regulations, the amendments themselves relate solely to matters of procedure. Requirements to keep records and make reports and regulations providing specific instructions as to where applications should be filed, how they should be signed and executed, the number of copies to be furnished, and the procedural steps which will be followed in connection with their review, do not have an effect on the environment. Like the amendments in Category 1., their function is to facilitate the orderly conduct of Commission business.

2.2.4 Background for 10 CFR 51.22(c)(10)

This CATX states:

> (10) Issuance of an amendment to a permit or license pursuant to parts 30, 31, 32, 33, 34, 35, 36, 39, 40, 50, 60, 61, 70 or part 72 of this chapter which (i) changes surety, insurance and/or indemnity requirements, or (ii) changes recordkeeping, reporting, or administrative procedures or requirements.

Discussion and Finding (49 FR 9352)

Issuance of an amendment to a permit or license to change surety, insurance and/or indemnity requirements or to change requirements relating to recordkeeping, reporting or other administrative procedures does not affect the scope or nature of the licensed activity. Although changes in surety, insurance and/or indemnity requirements affect the financial arrangements of licensees and have economic and social consequences, they do not alter the environmental impact of the licensed activities. Similarly, changes in recordkeeping and reporting requirements and other administrative procedures relating to the licensee's organization and management do not change the nature and the consequent environmental impact of the licensed activity. The function of these procedural and administrative changes is merely to facilitate the orderly conduct of the licensee's business and to insure that the information needed by the Commission to perform its regulatory functions is readily available.

2.2.5 Background for 10 CFR 51.22(c)(11)

This CATX states:

> (11) Issuance of amendments to licenses for fuel cycle plants and radioactive waste disposal sites and amendments to materials licenses identified in § 51.60(b)(1) which are administrative, organizational, or procedural in nature, or which result in a change in process operations or equipment, provided that (i) there is no significant change in the types or significant increase in the amounts of any effluents that may be released offsite, (ii) there is no significant increase in individual or cumulative occupational radiation exposure, (iii) there is no significant construction impact, and (iv) there is no significant increase in the potential for or consequences from radiological accidents.

Discussion and Finding (49 FR 9352)

Some requests for amendments to these types of licenses are administrative, organizational or procedural in nature or involve changes in process operations and equipment which do not result in any significant adverse incremental impacts to the environment from the licensed activity. Implementation of these minor and routine types of changes do not significantly alter the previously evaluated environmental impacts associated with the licensed operation, taking into account construction impacts, types and amounts of effluents released by the operation, occupational exposure of employees, or potential accidents. Furthermore, these amendments do not affect the scope or nature of the licensed activity.

For this CATX to apply, the license amendment must involve routine and minor types of changes that do not significantly alter the previously evaluated environmental impacts associated with the licensed operation, and the amendment must not affect the scope or nature of the licensed activity. This CATX has additional documentation requirements as described in Section 2.1

This CATX does not apply to the approval of decommissioning activities or decommissioning plans. This CATX may apply for certain decommissioning actions such as a change in facility status from operational to decommissioning, however, no additional decommissioning activities could be approved as this would constitute a change in the scope and nature of the licensed activity; only decommissioning activities authorized under the existing license would be permitted. For license actions to authorize decommissioning plans or additional decommissioning activities, the CATX listed in 10 CFR 51.22(c)(20), must be used, if appropriate, otherwise an EA should be prepared (see Section 2.2.7).

2.2.6 Background for 10 CFR 51.22(c)(12)

This CATX states:

> (12) Issuance of an amendment to a license pursuant to parts 50, 60, 61, 70, 72 or 75 of this chapter relating solely to safeguards matters (i.e., protection against sabotage or loss or diversion of special nuclear material) or issuance of an approval of a safeguards plan submitted pursuant to

parts 50, 70, 72, and 73 of this chapter, provided that the amendment or approval does not involve any significant construction impacts. These amendments and approvals are confined to (i) organizational and procedural matters, (ii) modifications to systems used for security and/ or materials accountability, (iii) administrative changes, and (iv) review and approval of transportation routes pursuant to 10 CFR 73.37.

Discussion and Finding (49 FR 9352)

These amendments are needed (1) to implement new safeguards regulations through incorporation of provisions into licenses, if requested and (2) to permit modifications to licensees' safeguards programs established under existing requirements.

With regard to route approvals, the requirement in 10 CFR 73.37(b)(7) for advance NRC approval of transportation routes applies only to spent fuel shipments and was included in the Commission's regulations in order to provide additional assurance that shipments containing spent fuel would be adequately protected against loss, diversion or sabotage. Before approving a particular transportation route, the NRC first makes a determination, on the basis of independently acquired information, that (1) details have been worked out for swift response by local law enforcement agencies, if requested, and (2) concrete details for NRC contingency planning for the route are adequate.

The Commission in NUREG-0170 ["Final Environmental Impact Statement on the Transportation of Radioactive Material by Air and Other Modes," (NRC, 1977)], *a generic environmental impact statement, considered the environmental impacts of the transportation of radioactive materials, including the transportation of those materials over routes approved for safeguards purposes, and concluded that such impacts are small. This generic environmental impact statement set out the NRC's views of the present (1977) and projected (1985) environmental impact of the transportation of radioactive material and provided documentation for the NRC determination that the environmental impacts, radiological as well as non-radiological, of both the normal transportation of radioactive materials and of the risk and consequent environmental impacts attendant on accidents involving radioactive material shipments were sufficiently small that shipments by all modes of transport should be allowed to continue and that no immediate changes to NRC regulations were needed. This report also concluded that the risks of theft or sabotage resulting in any significant radiological release are sufficiently small to constitute no major adverse impact on the environment. The Commission has examined the potential impacts set forth in NUREG-0170 and characterized as "small" and determined that they do not amount to a significant adverse impact.*

The Commission notes, however, that if special circumstances are shown to exist in connection with a particular shipment an environmental assessment or an environmental impact statement may be prepared for that shipment, and that as further review continues, this conclusion may be modified.

2.2.7 Background for 10 CFR 51.22(c)(14)

This CATX states:

(14) Issuance, amendment, or renewal of materials licenses issued pursuant to 10 CFR parts 30, 31, 32, 33, 34, 35, 36, 39, 40 or part 70 authorizing the following types of activities:

and encompasses 16 specific types of activities, as discussed below.

2.2.7.1 CATX 51.22(c)(14)(i)

This CATX states:

> (i) Distribution of radioactive material and devices or products containing radioactive material to general licensees and to persons exempt from licensing.

Discussion and Finding (49 FR 9352)

These licenses authorize persons to distribute radioactive materials and devices such as density gauges, level gauges, and other gauging devices to persons who are general licensees and to distribute products containing radioactive material such as watches, electron tubes, or smoke detectors to persons who are exempt from licensing. These licenses for distribution do not authorize processing or use of radioactive materials. There are no effluent releases or personnel exposures associated with the licensed activities. These distribution licenses presuppose ultimate use or possession of the radioactive materials under a general license or exemption established by regulation, which regulation, under 51.21, will require an environmental assessment addressing the environmental impacts of the generally licensed or exempted activities of the recipients of the materials. The radioactive material, devices and products that may be distributed pursuant to these licenses must meet the specific standards and requirements in the NRC regulations. At the time of issuance of the regulations authorizing distribution, the determination was made that subsequent exempt or generally licensed use or possession of the materials would not constitute a risk to the public health and safety.

2.2.7.2 CATX 51.22(c)(14)(ii)

This CATX states:

> (ii) Distribution of radiopharmaceuticals, generators, reagent kits and/ or sealed sources to persons licensed pursuant to 10 CFR 35.18.

Discussion and Finding (49 FR 9352)

These licenses for distribution do not authorize possession, use or processing of radioactive materials. There are no effluent releases or personnel exposures associated with the licensed activities.

2.2.7.3 CATX 51.22(c)(14)(iii)

This CATX states:

> (iii) Nuclear pharmacies.

Nuclear pharmacies purchase prepared radiopharmaceuticals, radioisotope generators, and reagent kits from manufacturers. They elute the generators and distribute the eluate as a prepared radiopharmaceutical or compound the eluate with reagent kits to make prepared radiopharmaceuticals. They dispense and distribute prepared radiopharmaceuticals to medical licensees in unit-dose or multi-dose forms. If the services of a nuclear pharmacy are not used, the medical licensee performs these functions in his own nuclear medicine laboratory. Due to the short half-life of medically useful isotopes, the radioactive wastes that nuclear pharmacies generate may be decayed to background levels in storage. Releases in effluents may be estimated at 5% of maximum permissible values. Due to the soft gamma emission of most medically useful isotopes and the use of personnel shielding devices, exposure to personnel may be conservatively estimated at 25% of the maximum permissible dose.

2.2.7.4 CATX 51.22(c)(14)(iv)

This CATX states:

(iv) Medical and veterinary.

Discussion and Finding (49 FR 9352)

NRC issues licenses to hospitals and to physicians authorizing use of radioactive materials in the diagnosis and treatment of patients. These licensed activities may include such activities as: receipt of radioactive material, preparation of radiopharmaceuticals from Mo-99/Tc-99m generators and reagent kits, administration of unsealed radiopharmaceuticals to patients for diagnostic or therapeutic purposes, the use of sealed sources for brachytherapy (i.e., radiation delivered from a short distance) and/or teletherapy (i.e., radiation delivered from a long distance), use of sealed sources contained in devices implanted in patients (e.g., nuclear-powered pacemakers), laboratory use of unsealed sources for performance of diagnostic tests or for tracer studies for research purposes, use of source material for shielding (e.g., as a component of a teletherapy unit or a linear accelerator), and the disposal of the authorized materials by holding for decay or by transfer to authorized recipients.

For the purposes of this discussion, medical licenses also include similar activities conducted by veterinarians for diagnosis or treatment of animals and laboratory use of unsealed sources for diagnostic tests as performed by clinical laboratories.

The environmental impact of these licensed activities is insignificant. In light of 10 CFR 20.107, radiation exposures of patients are not considered. The environmental impacts would be: occupational exposures estimated at less than 10% of the applicable limits; non-occupational exposures of members of the public who may have contact with these patients are generally minimal; releases to air and water or to sanitary sewerage (primarily as patient excreta) are of small quantity, or if of larger quantities, are short-lived. Effluent releases with the exception noted in 10 CFR 20.303(d) are estimated at less than 10% of the applicable limits.

2.2.7.5 CATX 51.22(c)(14)(v)

This CATX states:

> (v) Use of radioactive materials for research and development and for educational purposes.

Discussion and Finding (49 FR 9352)

These licenses authorize persons (e.g., academic institutions, industrial firms, and government agencies) to use sealed and/or unsealed sources of byproduct, source and special nuclear material for activities such as research and development (10 CFR 30.4(q)), educational purposes, classroom demonstrations, animal tracer studies, and tracer studies of materials and compounds. (Licenses to construct or operate nuclear research reactors are not materials licenses and therefore are not within the scope of this categorical exclusion.) This categorical exclusion does not encompass (a) processing or manufacturing, (b) performance of field studies in which licensed material is deliberately released directly into the environment for purposes of the study, or (c) use of radioactive tracers in field flood studies involving secondary and tertiary oil and gas recovery. As specified in 51.60(b)(1)(vi), applicants seeking licenses authorizing the use of tracers in field flood studies involving secondary and tertiary oil and gas recovery are required to submit environmental reports. In the case of other field studies in which licensed material is deliberately released directly into the environment for purposes of the study, environmental reports will be requested on a case-by-case basis as needed.

A typical facility is designed to minimize release of effluents to the environment. Remote handling equipment, personnel protective clothing, and shielding materials are standard equipment to minimize personnel exposures. A day-to-day radiation safety program provides for monitoring of personnel exposures, contamination levels, radiation levels, and effluent releases. Personnel exposures and effluent releases are estimated at less than 10 per cent of the limits of 10 CFR Part 20.

This CATX deals with the issuance, amendment, or renewal of materials licenses authorizing the use of radioactive materials for research and development and for educational purposes. This CATX does not apply to the five types of activities listed in Section 2.2.

In applying this CATX the following should be considered:

- Radioactive material used including its half-life, chemical characteristics, and solubility;
- Procedures to control and clean-up the radioactive material;
- Location, size, and length of study; and
- Ability to restrict access to study area.

If a research and development or academic institution application proposes to release to the environment radioactive materials that originated on-site (i.e., within the controlled property of the licensee), an EA is normally not needed and is covered under this categorical exclusion, provided:

- All releases originating on-site to the environment, such as air and liquid effluents, direct radiation from deposition of radioactive materials from the release (e.g., groundshine), comply with as low as reasonably achievable (ALARA) and 20 requirements.

- To assist in demonstrating compliance with the requirements of 10 CFR 20, the licensee should set ALARA goals for air effluents at a modest fraction of the values in Appendix B, Table 2, Columns 1 and 2, to 10 CFR 20.1001-20.2401. Experience indicates that values of about 10 millirems per year from all of the licensee's radioactive air effluents should be practicable for almost all materials facility licensees (see Regulatory Guide 8.37); therefore, as a first step toward demonstrating compliance with ALARA for radioactive air effluents, the licensee demonstrates that the nearest member of the general public receives no more than 10 millirems per year from all of the licensee's radioactive air effluents (i.e., licensee demonstrates it meets the requirements of 10 CFR 20.1101(d)).

- All releases on-site comply with all applicable decommissioning requirements (e.g., decommissioning recordkeeping requirements pursuant to 10 CFR 30.35(g)) and current decommissioning policies.

Documentation that supports the licensee's application as meeting the above criteria is sufficient to support why an EA is not needed. Research activities which may qualify for this CATX with an explanatory memorandum or other appropriate documentation include:

- Those that involve crops on small plots which are planted in a lined area and totally removed at the conclusion of the study and the study design prevents release to the environment and includes confirmatory analysis of the soil beneath the liner;

- Those that involve a small quantity of short-lived material which will decay to natural background by the conclusion of the study and the study design includes confirmatory analysis of background levels; and

- Those that involve tagging of animals and penning them to prevent their escape.

For license actions that cannot meet the above criteria, the Regions should coordinate with EPAB to determine whether an EA is needed. For example, an EA would be required for discrete sources released to the environment, which originated on-site, and which may not be recovered at the conclusion of the study or decommissioning. Examples of field studies that might require an EA include:

- Those that are not similar to normal routine research, development and educational activities;

- Those that deliberately release discrete sources to the environment;

- Those that release sources that may not be recovered; or

- Those that affect endangered species or historical/cultural resources.

2.2.7.6 CATX 51.22(c)(14)(vi)

This CATX states:

(vi) Industrial radiography.

Discussion and Finding (49 FR 9352)

Gamma radiation sources (primarily iridium-192 and cobalt-60) are used for non-destructive testing of materials throughout the United States. The sources used are metallic and are encapsulated in a stainless steel capsule. Therefore, during ordinary use it is not expected that there will be releases of radioactive material to the environment. The radiation exposure during routine use of sources in industrial radiography is well within NRC limits for occupational exposure. The average exposure per individual radiographer is less than 0.4 rem per year, which is less than 10% of the permissible exposure.

2.2.7.7 CATX 51.22(c)(14)(vii)

This CATX states:

> (vii) Irradiators.

Discussion and Finding (49 FR 9352)

These devices are used for a variety of purposes in research and industry to expose products to large amounts of radiation. Typical uses include sterilization or microbiological reduction in medical and pharmaceutical supplies and insect eradication through sterile male release programs. Irradiators usually contain from a few hundred curies to megacuries of radioactive material, principally cobalt 60. The radioactive material is contained in sealed sources. Product irradiation occurs within areas to which access is controlled and which are shielded to protect both operating personnel and the environment.

Personnel exposures during use of these devices are less than 5% of the limits in 10 CFR Part 20. There are no effluent releases resulting from operation of irradiators.

2.2.7.8 CATX 51.22(c)(14)(viii)

This CATX states:

> (viii) Use of sealed sources and use of gauging devices, analytical instruments and other devices containing sealed sources.

Discussion and Finding (49 FR 9352)

Sealed sources used by licensees are usually singularly or doubly encapsulated depending on activity in stainless steel. Therefore, in ordinary use it is not expected that the use of sealed sources will result in the release of radioactive material to the environment. Sealed sources used by licensees are usually required to undergo rigorous prototype testing to ensure that the likelihood of a substantial release of radioactive material to the environment during abnormal use of sealed sources is unlikely.

Gauging devices used to measure thickness, density, and level of materials contain sealed sources, usually cesium-137 and strontium-90, which are encapsulated so that there is no leakage during use. The devices provide shielding such that radiation levels external to the devices are on the order of a few milliroentgens per hour. Other devices include gas chromatographs with millicurie quantities of nickel-63 or hydrogen-3, analytical devices such as X-ray fluorescence analyzers with sealed sources containing a variety of radioisotopes, instrument calibration devices containing millicurie to curie quantities of cesium-137 and cobalt-60, and soil-density gauges which contain millicurie quantities of cesium-137 and americium-241 neutron sources.

Personnel exposure from use of these devices is less than 5% of the limits in 10 CFR Part 20. There are no effluents associated with the use of devices containing sealed sources.

2.2.7.9 CATX 51.22(c)(14)(ix)

This CATX states:

> (ix) Use of uranium as shielding material in containers or devices.

Discussion and Finding (49 FR 9352)

These licenses for possession and use of uranium for shielding are a non-nuclear use of radioactive materials. Because of its high density, uranium is excellent as shielding material. Depleted uranium has very low specific activity and the corresponding low radiation levels emitted make it very unlikely that any individual will receive a radiation dose in excess of 5% of maximum permissible dose specified in Part 20. In addition, because of its physical and chemical properties, there should be no release of radioactive material to the environment during normal use of depleted uranium as shielding and very limited release during abnormal conditions.

2.2.7.10 CATX 51.22(c)(14)(x)

This CATX states:

> (x) Possession of radioactive material incident to performing services such as installation, maintenance, leak tests and calibration.

Discussion and Finding (49 FR 9352)

These licenses only authorize the possession of radioactive material incident to performing services either at the customer's facility or at the licensee's facility. Generally the activity involves the use of sealed sources only. Since service licenses involved very little actual possession and use of radioactive material, personnel exposure from performing the services should be less than 5% of the limits in 10 CFR Part 20 and there should be no effluent releases.

2.2.7.11 CATX 51.22(c)(14)(xi)

This CATX states:

> (xi) Use of sealed sources and/or radioactive tracers in well-logging procedures.

Discussion and Finding (49 FR 9352)

During the past 20 years in which the NRC and its predecessor agency, the AEC, have been regulating the use of sealed radioactive sources and short-lived radioactive tracers in well logging, there have been approximately 89 incidents in which well-logging sources have been forced to be abandoned in wells. A risk analysis prepared by the NRC staff shows only a small radiological risk to the public health and safety from the potential release of radioactive material due to long term corrosion or damage from drilling into sources that have been abandoned. In addition, routine safety measures, such as those described below, also protect against significant environmental impacts from well-logging activities.

Well drilling permits require that gas and oil wells be cased to below potable water aquifers to prevent cross contamination from brine, oil and gas normally associated with wells. This requirement also serves to preclude contamination of potable water aquifers when radioactive materials are used in these cased wells. In the event a source becomes irretrievable during a well-logging operation, safety requirements are imposed to minimize the escape of radioactivity from the source and the surrounding areas. These requirements include: (1) Sealing the source in place with a cement plug to immobilize it and to preclude abrasion and corrosion; (2) setting a deflection device (whipstock) at the top of the cement plug to deflect a drill away from the general area of the source in the event of an inadvertent future drilling; (3) mounting a permanent identification plaque at the surface of the well to alert anyone planning to enter the well to the existence of a source downhole; and (4) requiring notification to be placed in pertinent land records maintained by State oil and gas regulatory agencies to alert against re-drilling the well. In addition, the construction of the source itself minimizes the possibility of releases and migration of radioactive material. Source capsules are always doubly encapsulated and fabricated of stainless steel or other corrosion resistant material. The radioactive material is in the form of a very low solubility compound. The sources are enclosed in a logging tool made of steel which provides additional protection.

The radioactive materials used as tracers in well logging have short half-lives and the quantities involved are small--in the low millicurie range. The use of these tracers does not present any environmental impact because of the small quantities which decay to innocuous radioactivity levels in short periods of time.

Additional guidance can be found in NUREG-1556, Vol. 14, "Consolidated Guidance About Materials Licenses: Program-Specific Guidance About Well Logging, Tracer, and Field Flood Study Licenses" (NRC, June 2000b).

2.2.7.12 CATX 51.22(c)(14)(xii)

This CATX states:

> (xii) Acceptance of packaged radioactive wastes from others for transfer to licensed land burial facilities provided the interim storage period for any package does not exceed 180 days and the total possession limit for all packages held in interim storage at the same time does not exceed 50 curies.

These licenses authorize the acceptance of radioactive waste in packages that meet all governmental regulations for transport of radioactive materials. The packaged radioactive material is stored temporarily until a sufficient number of packages is accumulated for shipment to licensed land burial sites.

In general, these activities are analogous to the transport carried out by common and contract carriers, which are exempt from NRC license requirements. Packages are not permitted to be opened although over-packaging may be carried out in the event defective packaging is received. There are no routine releases of radioactive effluents. Safety requirements for the storage facility include protection against unauthorized entry, fire resistant buildings and packages, fire detection and suppression capability, radiation monitoring equipment and operating and emergency procedures. By limiting the total radioactivity in storage at any one time to a maximum of 50 curies and by limiting the storage period for any package to a maximum of 180 days, the chances of significant releases of radioactivity or excess exposure of personnel in the event of accident conditions, such as a fire, are minimal.

2.2.7.13 CATX 51.22(c)(14)(xiii)

This CATX states:

> (xiii) Manufacturing or processing of source, byproduct, or special nuclear materials for distribution to other licensees, except processing of source material for extraction of rare earth and other metals.

Discussion and Finding (49 FR 9352)

Manufacturing or processing of source, byproduct, or special nuclear materials for distribution to other licensees consists of approximately 234 NRC licensees at the present time. Under these licenses, persons manufacture radiopharmaceuticals, labeled compounds for research purposes and sealed sources for use in gauging and analytical equipment. Other licensees in this category use and handle radioactive materials in solid form to manufacture sealed sources, e.g., radiography devices, or use and handle already sealed sources by incorporating the sources into devices used for gauging purposes.

In 1978, licensees in this category had an average dose of 0.45 rem for persons with measurable exposure and an average dose of 0.21 rem for all persons monitored. The collective dose for this category of licensees was 3,280 man-rems. The potential impact, therefore, is very small, less than one calculated health effect. Ninety-eight percent of the facilities had releases in air of less than one percent of the maximum permissible concentrations in 10 CFR Part 20. The largest release reported was approximately 12 percent of the maximum permissible concentrations. Releases of liquid wastes were well within the limits in NRC regulations.

Operations where source material is processed for extraction of rare earth or other metals may involve generation of large volumes of waste containing low levels of radioactive material. The storage and ultimate disposal of this waste may have significant environmental impact. Therefore, these types of operations are not listed as a categorical exclusion.

2.2.7.14 CATX 51.22(c)(14)(xiv)

This CATX states:

>   (xiv) Nuclear laundries.

Discussion and Finding (49 FR 9352)

Nuclear laundries receive slightly contaminated clothing from nuclear facilities and provide decontamination services. The "clean" garments are then returned to the customer. As of August 31, 1981, there were four NRC licensees in this category. The quantities of radioactive material involved are small, usually a few millicuries of radioactive material. In 1978, three of the four licensed laundries reported an average dose of 0.22 rem for persons with measurable exposure and a collective dose of 1 rem. The small amount of activity used by those licensees is disposed of in accordance with NRC regulations.

2.2.7.15 CATX 51.22(c)(14)(xv)

This CATX states:

>   (xv) Possession, manufacturing, processing, shipment, testing, or other use of depleted uranium military munitions.

Discussion and Finding (49 FR 9352)

Possession, manufacturing, processing, shipment, testing or other use of depleted uranium munitions, e.g., bullets and other projectiles, includes about 10 licenses held by U.S. military organizations and less than 10 licensees involved with the manufacturing process. The military tests involve the use of low specific activity depleted uranium ($3.6 \times 10\ 7$ curies/gram) as metal alloy penetrators (rods) which vary in weight from a few grams to less than 10 kilograms. These rods are propelled at high velocities against metal targets such as armor plate. Testing of these munitions is carried out at remote desert locations on military reservations, in constructed enclosures, or over deep ocean waters. Any materials released to the environment are of low radioactive content, are highly dispersed, and are of chemical and physical form which is not readily incorporated into flora or fauna. Thus, radioactive releases to the environment which could affect human, animal or plant life from testing at any of the locations are negligible and occupational exposures from handling depleted uranium are so low that personnel monitoring is not required. Additionally, since the penetrators tested do not explode, cratering or other defacing of the environment is not experienced. The military also transports and stores depleted uranium munitions for war-readiness posture. Because the munitions are transported and stored in sealed containers as solid metal in nondispersible form, there is negligible environmental impact associated with such transportation and storage.

Manufacturers of depleted uranium munitions are also included here for the sake of completeness, although manufacturers are excluded in section (xiii) of Category 14.

2.2.7.16 CATX 51.22(c)(14)(xvi)

This CATX states:

> (xvi) Any use of source, byproduct, or special nuclear material not listed above which involves quantities and forms of source, byproduct, or special nuclear material similar to those listed in paragraphs (c)(14) (i) through (xv) of this section (Category 14).

Discussion and Finding (49 FR 9352)

It has been the Commission's experience in the past that additional environmentally insignificant materials licensing actions occasionally arise. These cases involve uses of source, byproduct or special nuclear material in quantities and form similar to those categorically excluded in sections (i)-(xv) of Category 14 and, therefore, have insignificant environmental impacts. By categorically excluding actions of this type, the Commission will avoid the unnecessary expenditure of scarce resources in preparing environmental assessments for those few environmentally insignificant cases not separately identified as the subject of a specific categorical exclusion. The Commission anticipates that considerably less than 1% of its licensing actions in the nuclear materials area would fit within this category.

License actions not specifically listed in 10 CFR 51.22(c)(14) for which staff believes fit the criteria for a CATX will require a TAR to EPAB. To expedite the processing of the TAR, the Regions should perform an initial technical assessment, to be enclosed with the TAR, to justify why the licensing action qualifies for categorical exclusion under 10 CFR 51.22(c)(14)(xvi). Examples of the specific type of information that should be submitted to EPAB to assist the staff in making its determination can be found in Appendix I of NUREG-1556, Vol. 20, "Consolidated Guidance About Materials Licenses: Guidance About Administrative Licensing Procedures" (NRC, 2000a). When a TAR is received from the Region, EPAB will review the documentation and determine if the action qualifies for a categorical exclusion. EPAB will then provide a memorandum to the Region, documenting the results that need to be included in the docket. This CATX has additional documentation requirements as described in Section 2.1.

2.2.8 Background for 10 CFR 51.22(c)(17)

This CATX states:

> (17) Issuance of an amendment to a permit or license pursuant to parts 30, 40, 50 or part 70 of this chapter which deletes any limiting condition of operation or monitoring requirement based on or applicable to any matter subject to the provisions of the Federal Water Pollution Control Act.

Discussion and Finding (49 FR 9352)

The NRC no longer has a role setting conditions relating to nonradiological discharges of pollutants into aquatic bodies or establishing requirements for aquatic monitoring where an NPDES permit is in effect. Instead, EPA, and those states to whom permitting authority has been delegated, have exclusive responsibility for regulating nonradiological pollutant discharges through the NPDES permit system. The NRC's role in the water quality area is limited to regulating radiological discharges into aquatic bodies and NEPA matters such as weighing aquatic impacts in the NEPA analysis which NRC is required to make before reaching a major Federal licensing decision.

2.2.9 Background for 10 CFR 51.22(c)(19)

This CATX states:

> (19) Issuance, amendment, modification, or renewal of a certificate of compliance of gaseous diffusion enrichment facilities pursuant to 10 CFR part 76.

Discussion and Finding (59 FR 48944)

The two plants to be regulated by this rule have already been subject to evaluation in accordance with NEPA. The NRC has reviewed these documents, as well as environmental reports prepared by DOE for both facilities in 1992 and environmental audits prepared by DOE prior to turning operation of the Facilities over to the Corporation in 1993. The promulgation of a rule governing these plants, and their subsequent regulation by the NRC, will not result in any environmental impacts beyond those previously considered by DOE in its environmental reviews and which currently exist or would be expected to continue absent NRC regulatory oversight.

Similarly, subsequent certificates of compliance including amendments, modification, and renewals issued pursuant to this part will consist of findings of compliance with 10 CFR Part 76. Therefore, these actions will not result in any significant new environmental impacts. The regulations require that the Corporation submit information for use by NRC in preparing an environmental assessment for certification applications addressing areas where the facilities are not in compliance with the requirements of Part 76.

2.2.10 Background for 10 CFR 51.22(c)(20)

This CATX states:

> (20) Decommissioning of sites where licensed operations have been limited to the use of— (i) Small quantities of short-lived radioactive materials; or (ii) Radioactive materials in sealed sources, provided there is no evidence of leakage of radioactive material from these sealed sources.

Discussion and Finding (62 FR39058)

The Generic Environmental Impact Statement (GEIS) ["Generic Environmental Impact Statement in Support of Rulemaking on Radiological Criteria for License Termination of NRC –Licensed Nuclear Facilities," NUREG-1496 (NRC, 1997)] *prepared by the Commission on this rulemaking evaluates the environmental impacts associated with the remediation of several types of NRC-licensed facilities to a range of residual radioactivity levels. The Commission believes that the generic analysis will encompass the impacts that will occur in most Commission decisions to decommission an individual site where the licensee proposes to release the site for unrestricted use. Therefore, the Commission plans to rely on the GEIS to satisfy its obligations under the National Environmental Policy Act regarding individual decommissioning decisions that meet the 0.25 mSv/y (25 mrem/y) criterion for unrestricted use.*

However, the Commission will still initiate an environmental assessment regarding any particular site, for which a categorical exclusion is not applicable, to determine if the generic analysis encompasses the range of environmental impacts at that particular site. The GEIS indicates that the decommissioning for certain classes of licensees (e.g., licensees using only sealed sources) will not individually or cumulatively have a significant effect on the human environment.

The GEIS (NRC, 1997), Section 3.2.1, further noted that a large majority of NRC's 7000 materials licensees use either sealed radioactive sources or small amounts of short-lived radioactive materials in their business operations. Typically, these facilities can be categorized in the following manner:

1.   A sealed source is defined in 10 CFR 30 as any byproduct material that is encased in a capsule designed to prevent leakage or escape of the by product material. Sealed source users, licensed pursuant to 10 CFR Part 30, include medical users of sealed sources (teletherapy, brachytherapy), users of industrial gauges, well loggers, radiographers, and irradiators. Nuclides contained in the capsules and used by sealed source users include Co-60, Cs-137, I-125, Ir-192, Sr-90, and Am-241. The sealed sources are designed and tested according to the requirements of industrial standards and radiation safety criteria set out in the regulations to prevent leakage.

      As a result of the nature of the sealed source design, testing, and operation, it is expected that contamination of facility structures and soils would not result from routine operations.

      Recent experience indicates that the frequency of leakage of sealed sources is very low. Leaking sources are taken out of service and returned to another specific licensee (typically the manufacturer) for disposal. Sealed source contamination would most likely be contained within the device or otherwise localized, and remediation would be straightforward and localized. When operations using the sealed source cease, the sealed source would be returned to a specific licensee authorized to possess the source or sent to licensed disposal site for proper disposal. It is expected that decontamination of the building or soils would not be needed. Currently, 10 CFR 30.36 requires that sealed source licensees properly dispose of the source, submit NRC Form 314, and either conduct a radiation survey or demonstrate that the premises are suitable for license termination by other means.

2.   Licensees using short-lived byproduct radionuclides are licensed pursuant to 10 CFR Part 30 and use short-lived nuclides for specific reasons, primarily in the area of medical diagnostics. Short-lived nuclides licensed for such use include Tc-99m, I-131, and I-123.

      The nature of operations using short-lived nuclides, makes the contamination of facility structures and soils unlikely. Contamination (if any) would likely be confined to localized areas in buildings. Any such contamination would be diminished by radioactive decay, and no long-term contamination would remain after license termination. Cleanup would be straightforward and localized. The predominant means for decommissioning of facilities that use short-lived nuclides is "decay-in-storage." In terminating the license, the licensee follows the same procedure required under 10 CFR

30.36 as noted above for sealed sources; i.e., any byproduct material is properly disposed of, NRC Form 314 is submitted indicating disposition of any licensed material, and either a radiation survey is conducted or there is a demonstration that the premises are suitable for license termination by other means (e.g., by calculation of the reduction in activity by radioactive decay). Based on use of "decay-in-storage" for the short-lived nuclides, and the time involved in submitting the information necessary to terminate a license, it is expected that licensed material would reach sufficiently low levels such that decontamination of the building or of soils would not be needed.

This CATX deals with the decommissioning of sites where licensed operations have been limited to the use of small quantities of short-lived material or radioactive materials in sealed sources. Typically this CATX will be limited to the application of Group I decommissioning actions as defined in NUREG-1757, Vol. 1, "Consolidated NMSS Decommissioning Guidance" (NRC, 2003).

2.3   Consultations

CATX actions do not require EAs or EISs; they are "excluded" from more detailed levels of NEPA analysis. However, these actions are not excluded from other Federal, State, or local environmental laws and regulations. Therefore, the licensing PM may need to conduct additional analyses, consult with other agencies, carry out public participation activities, and prepare documentation under other applicable laws even though the proposed action qualifies for a CATX (e.g., a project to decommission a building included in or eligible for the National Register of Historic Places). However, in most cases, external environmental experts and agencies with jurisdiction by law or expertise such as the FWS or appropriate SHPO will not need to be consulted for licensing actions that qualify for a CATX as these actions typically "will not affect" listed species/critical habitat and/or are not the types of activities "with a potential" to affect historic and cultural resources. Appendix D provides a detailed procedure to follow in conducting consultations.

2.4   Public Participation in CATXs

Generally, determining whether an action is a CATX requires no public participation, but if an individual or group expresses interest in the project's environmental effects, they should be kept informed of the CATX review and a copy of the completed CATX checklist (Appendix B) or other documentation should be part of the publically available information documenting the NRC decision.

2.5   The Environmental Checklist

The licensing PM may use the checklist provided in Appendix B to document the CATX and the considerations in using the applicable CATX. Instructions for completing the checklist are also presented in Appendix B.

The licensing PM should be aware of special circumstances in which the CATX may not apply. In general, if the action does not fall within the discussions presented above or if other special circumstances (e.g., unique, controversial, precedent setting, etc.) are present, the licensing PM should prepare an EA. Additionally, actions which are potentially significant, as discussed in Section 3.4.6.3, should generally be considered a special circumstance and an EA/EIS should be prepared.

The checklist consists of one question about the applicability of the selected CATX and four questions about the likelihood that a particular kind of environmental consequence will result from the proposed action. The licensing PM should consult with technical staff and EPAB, as necessary.

Based on internal review, external review (where appropriate), and research, check "YES," "NO," or "NEED DATA" for each question. Attach documentation as needed to support the answer. If the "NEED DATA" box is checked, the licensing PM may consult with an environmental PM about what data is needed and/or how to get it. Appendix B provides details to consider when completing the checklist.

The checklist is not complete until all "NEED DATA" issues have been resolved and all blocks are checked either "YES" or "NO." Checking a single block to "YES" does not necessarily mean that an EA must be prepared; it may be possible to resolve the "YES" answer in another way (e.g., additional technical documentation).

Resolve all "NEED DATA" issues and complete the checklist, attaching all supporting documentation. In the "Conclusions" section, check the box corresponding to the conclusion reached. The checklist is now complete. The checklist becomes part of the licensing file and can be made available to the public and other potential review agencies upon request. In some cases, it may be necessary to consult with OGC to respond adequately to the questions in the checklist.

The CATX checklist is a guide. The licensing PM should take appropriate steps to carry out the conclusions reached. If the conclusion is that further review is needed, the licensing PM should ensure this review happens. If the conclusion is that a CATX is not warranted, the licensing PM should ensure the appropriate level of analysis and documentation is carried out and initiate preparation of an EA. For any activity related to an EIS, the responsibility for the environmental review should be transferred to EPAB.

2.6   References

NRC (U.S. Nuclear Regulatory Commission), 1977. "Final Environmental Impact Statement on the Transportation of Radioactive Material by Air and Other Modes." NUREG-0170. U.S. Nuclear Regulatory Commission, Washington, D.C. December.

NRC, 1984. "SECY-83-286 - Revision to 10 CFR Part 51 and Related Conforming Amendments - Implementation of CEQ NEPA Regulations." Memorandum from Chilk to Dircks. U.S. Nuclear Regulatory Commission, Washington, D.C. February 28.

NRC, 1997. "Generic Environmental Impact Statement in Support of Rulemaking on Radiological Criteria for License Termination of NRC-Licensed Nuclear Facilities." NUREG-1496. U.S. Nuclear Regulatory Commission, Washington, D.C. July.

NRC, 2000a. "Consolidated Guidance About Materials Licenses: Guidance About Administrative Licensing Procedures." NUREG-1556, Volume 20. U.S. Nuclear Regulatory Commission, Washington, D.C. June.

NRC, 2000b. "Consolidated Guidance About Materials Licenses: Program-Specific Guidance About Well Logging, Tracer, and Field Flood Study Licenses." NUREG–1556, Volume 14. U.S. Nuclear Regulatory Commission, Washington, D.C. June.

NRC, 2001. "Regulations Handbook." NUREG/BR–0053, Revision 5. U.S. Nuclear Regulatory Commission, Washington, D.C. March.

NRC, 2003. "Consolidated NMSS Decommissioning Guidance." NUREG–1757, Volume 1, Revision 1. U.S. Nuclear Regulatory Commission, Washington, D.C. September.

PAGE INTENTIONALLY BLANK

3 PREPARING AN ENVIRONMENTAL ASSESSMENT

An EA must be prepared for proposed actions that are not:

- Exempt from NEPA (this does not mean licensing exemptions);

- Categorically excluded (10 CFR 51.22);

- Covered in an existing EIS or other environmental analysis; or

- Required to have an EIS prepared (10 CFR 51.21).

An EA may be prepared for any action to assist in planning and decision making (40 CFR 1501.3), but should be initiated as early in the process as possible after the license application submittal, license amendment request, or rulemaking action. The EA should provide sufficient evidence and analysis of impacts to support a determination of a finding of no significant impacts (i.e., FONSI). If an EA does not result in a FONSI, then the potential impacts from the proposed activities require the preparation of an EIS. The EA process is an interdisciplinary review of proposed actions and their impacts on all affected resources. The EA may also identify and develop appropriate mitigation measures, which may result in a mitigated FONSI. The EA and any related FONSI are made available to the public. In cases when an EIS is found to be necessary, any research completed during EA preparation can be used in the preparation of the EIS. If the action under review is certain to result in significant impacts, the environmental review to support the action should move directly to an EIS. As described in Chapter 1, responsibility for completing the EIS should be transferred to EPAB.

The EA process need not be extraordinarily time consuming or complicated. The level of assessment should be commensurate with the anticipated impacts and the degree of public concern. EAs are prepared by the licensing PM or rulemaking task leader responsible for the action associated with the EA along with the assistance of technical staff. Staff from EPAB (environmental PM) review all EAs, except for Certificate of Compliance rulemakings, prepared by NMSS staff[5] and will be available to assist with determinations on whether an action will require an EA or EIS, especially for areas where policy is being developed. Figure 2 provides an overview of the EA process.

3.1   Environmental Assessment Development

Following the acceptance review, as discussed in Section 1.3.3, the licensing PM and other necessary technical reviewers should develop a preliminary draft of the EA. This effort assists with identification of missing and unclear information, facilitates the preparation of requests to the applicant/licensee for additional information (RAI), and streamlines the EA development. RAI is a term applied to additional necessary information (clarifications and questions) requested of the applicant/licensee to complete the environmental and safety review. To streamline the process, the NMSS goal is to focus the content of the RAIs to that additional information necessary to support a regulatory decision. Preparation of a

---

[5]In the future, after Headquarters and Regional staff gain more experience preparing EAs consistent with this guidance, only "complex" EAs will need to sent to EPAB. However, if requested, EPAB will review "simple" EAs. EPAB will issue a memorandum notifying the staff of changes in the review requirements for EAs.

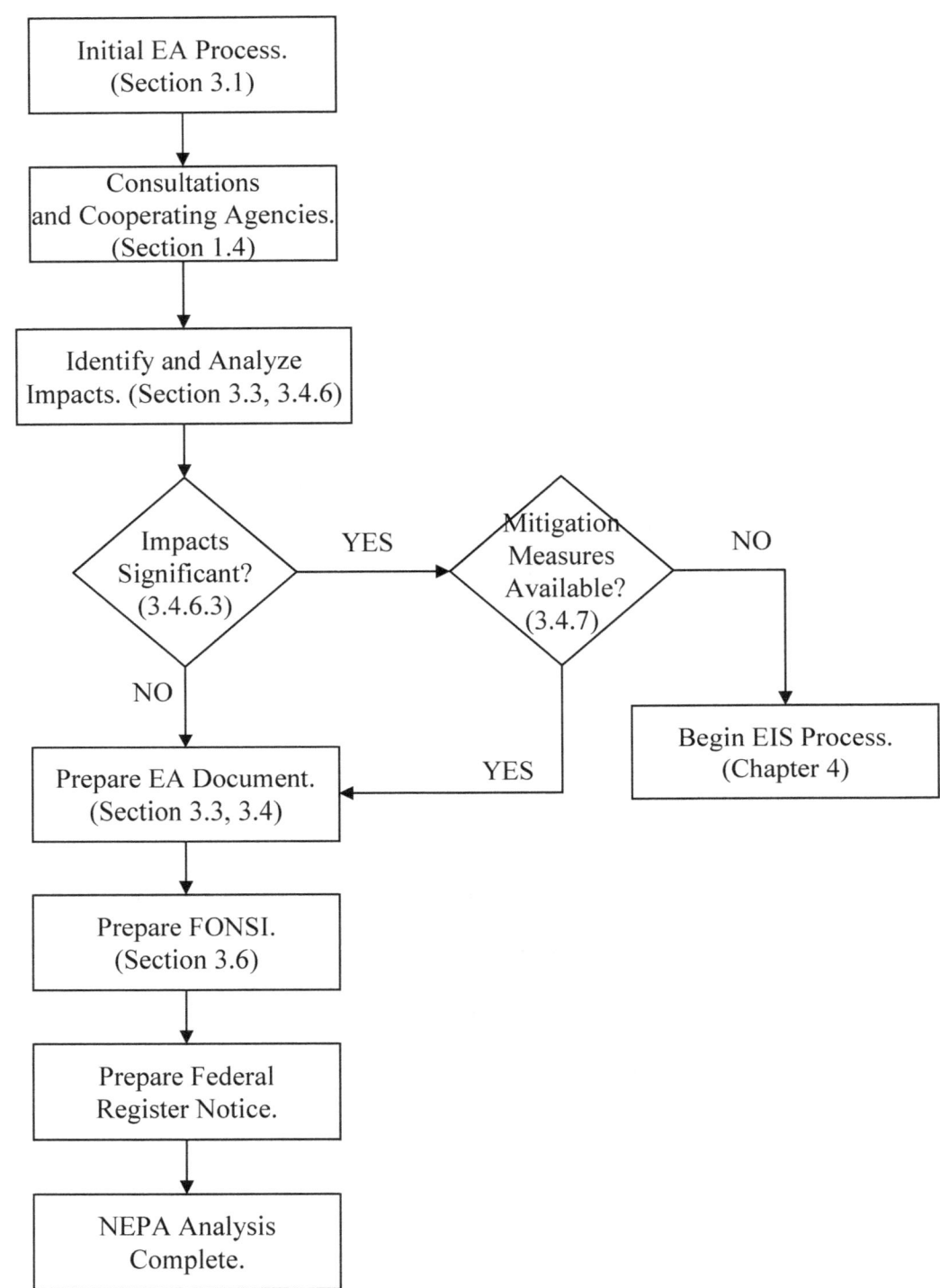

Figure 2: Major steps in the EA process.

preliminary draft EA ensures that the necessary information is being requested. The licensing PM should consult with EPAB to determine any recent policy changes for environmental reviews that might impact the RAIs. In evaluating the applicant's environmental information, the licensing PM and other technical reviewers should identify and evaluate the quality assurance measures taken by the applicant in collecting and analyzing data. Quality assurance measures, including verification and validation, are also evaluated where computer models have been used to predict environmental consequences of the proposed actions.

Related generic and site-specific EAs and/or EISs should be reviewed to determine if there is a potential for tiering (Section 1.6, *Utilizing Existing Environmental Analyses*). Attention should be given to the bounding conditions (both environmental and non-environmental) and related assumptions of these previous analyses to determine if they apply to the new proposed action. This comparison and determination should be briefly described in the EA and, for future generic use, may be documented in greater detail separate from the EA (e.g., response letter to person/organization providing comments). Applicable portions of existing EAs and/or EISs should be incorporated by reference to shorten the length of the EA.

An example of tiering off a GEIS is provided in Appendix A. This appendix contains a checklist that the licensing PM can use to determine whether it is appropriate to tier off the GEIS developed in support of the License Termination Rule (NRC, 1997). This checklist should only be used for sites being released for unrestricted use and is related only to dose assessments (i.e., nonradiological impacts must still be considered).

3.2   EA Format and Technical Content

At a minimum, an EA is required (10 CFR 51.30) to include a brief discussion of (i) the proposed action, (ii) the need for the proposed action, (iii) alternatives as required by Section 102(2)(E) of NEPA, (iv) the environmental impacts of the proposed action and alternatives, and (v) a list of agencies and persons consulted, and (vi) identification of sources used. Program-specific guidance may identify additional format and content requirements or options.

Due to the wide variation in the types of NMSS licensing actions, two different formats for preparing an EA are presented below, simple and complex. Simple licensing actions can include minor changes to existing facilities or administrative type actions which do not qualify for categorical exclusion and are discussed further in Section 3.3. EAs for these simple licensing actions should typically not exceed 5 pages. Complex licensing actions can include major changes to existing facilities (e.g., significant new construction), construction of new facilities, or approval of decommissioning plans involving major disturbances to the environment, and are discussed further in Section 3.4. EAs for these complex licensing actions should be consistent with CEQ guidance which suggests a 10-15 page limit. CEQ has also noted that lengthy EAs may be an indication that an EIS is needed (CEQ, 1981).

Rulemaking actions will generally use the simple format for preparing an EA, though more complicated rulemakings may use the complex format.

3.3   EA for Simple Licensing Actions

As discussed in Section 3.2, an EA is required to contain the following items.

Introduction

Though not required, an "Introduction" is usually helpful to orient the audience by providing an overview and should include a brief description of the proposed action, how and when the proposed action was submitted (e.g., license application, license amendment), and by whom the proposed action was submitted. Any unique terms and phrases (e.g., cask, sealed-source, restricted release) should be briefly defined as they are presented. A brief discussion of the relevant NRC regulations should be included. Also, other relevant documents (e.g., another agency's EA or a previous NRC review) should be described in relation to the proposed action and should be clearly referenced.

The Proposed Action

This section should describe the proposed action in more detail and reference the pertinent license application. Also, planned activities/phases, location, and duration of the proposed action should be described. Relevant and brief descriptions of proposed activities expected to result in impacts should be described in enough detail to support the environmental impacts discussion. Also, the description of the proposed action should describe any mitigation measures that have been incorporated into the action to avoid environmental impacts (i.e., measures the licensee has committed to doing as a part of the proposed action). The proposed action is what the applicant/licensee is proposing to do or accomplish with the license amendment. The proposed action is not NRC approval of the license/amendment request. Also, other relevant documents (e.g., another agency's EA or a previous NRC review) should be described in relation to the proposed action and should be clearly referenced.

The Need for the Proposed Action

This section should discuss the applicant's motivation for submitting the application to the NRC. For example, does the requested exemption or amendment provide some benefit to the applicant if granted? How would the applicant be affected if the application was not approved? The need should not be described as a justification for the proposed action over the alternatives. Also, the need should not be stated as NRC approval of a license request.

Environmental Impacts of the Proposed Action

This section should describe how the environmental resource (e.g., land or water) is used, how the resource would be affected by the proposed change (e.g., the construction of a building, change in the amount of water taken in by the facility, record keeping or reporting requirements, etc.), and the significance of the relationship between the environmental resource and proposed change. The EA should include an assessment of the direct, indirect, and cumulative impacts of the proposed action. For example, air quality (the environmental resource) would be affected by a release of radioactivity from increased facility effluents and the significance of the release would depend on the types and amounts of the emission. In this example, an appropriate question would be whether the emission for the radionuclide is above the regulatory limits or is a small fraction of the regulatory limits. The section should include an evaluation of radiological and nonradiological impacts. The impacts section should also describe the potential impacts to cultural or historic resources and threatened and endangered species or critical habitat (see Appendix D for more information on consultation requirements). It should clearly state which resources are affected by the proposed action. Likewise, it should clearly state no environmental resources are affected, if that is the case. Section 3.4.6 provides additional information for discussing environmental impacts.

Environmental Impacts of the Alternatives to the Proposed Action

NEPA requires NRC to consider alternatives in the preparation of all EAs whenever the following two conditions are present: (i) there is some identifiable environmental impact from the proposed action and (ii) the objective of the proposed action can be achieved in one of two or more ways that will have differing impacts on the environment. This section of the EA should discuss alternatives to the proposed action and the environmental impact of the alternatives. The fact that the EA involves a finding of no significant impact (FONSI) does not provide relief from considering alternatives. As long as there is some identifiable impact on the environment from the proposed action, the EA should consider alternatives.

For those actions involving a very small impact, it is reasonable to consider a very limited range of alternatives. In fact, in several decisions, the courts have stressed that the range of alternatives an agency must consider in an EA decreases as the environmental impact of the proposed action becomes less and less substantial. However, no court has held that an agency is excused from considering alternatives if the agency has made a FONSI, and, in fact, considering alternatives is independent of the question of environmental impact.

At a minimum, the no-action alternative must be addressed. The no-action alternative is a discussion of the results from a lack of action (i.e., status quo or the existing state). For example, if the proposed action is the clean-up of a site for unrestricted use, then the no-action alternative is to continue to keep the material licensed and on site, without disposal. More specific guidance on alternatives for an EIS is provided in Section 5.2.

A insignificant impact does not equate to no impact. Therefore, if an even less harmful alternative is feasible, then it needs to be considered. If the environmental impact of a proposed action is zero (e.g., administrative type actions), there is no meaningful alternative to be considered because there is no use of natural resources associated with the action. In those cases involving no environmental impact, it is reasonable to limit the discussion of alternatives to consideration of the no-action alternative. If the "no-action" alternative is the only alternative examined, the alternatives section may contain the following statement, if appropriate:

> "As an alternative to the proposed action, the staff considered denial of the proposed action (i.e., this is the "no-action alternative"). Denial of the application would result in no change in current environmental impacts. The environmental impacts of the proposed action and the alternative action are similar."

Section 3.4.6 provides additional information for discussing environmental impacts.

Agencies and Persons Consulted

This section of the EA should list Federal and State agencies and persons consulted. The licensing PM should consult with the affected State just prior to issuing the final EA and should solicit comments on the environmental impact of the proposed action and any other comments the State may have (NRC,

1994). Additionally, the licensing PM is responsible for ensuring that other appropriate agencies are contacted if an action may involve some impact on the natural or physical environment and these consultations should be initiated early in the development of the EA. Appendix D provides a detailed description for the licensing PM to follow for consultations:

- With the appropriate State official;
- Required under Section 7 of the Endangered Species Act; and
- Required under Section 106 of the National Historic Preservation Act.

All consultations should be briefly documented in the EA and should contain (i) the name of the agency or person contacted (consulted), (ii) the date and purpose of the consultation, (iii) a brief summary of the views or comments expressed and the staff's resolution, and (iv) references to publicly available documents containing additional information, as applicable.

For licensing actions that do not affect endangered or threatened species or do not have the potential to cause effects on historic properties the following statement should be considered:

"NRC staff have determined that the proposed action will not affect listed species or critical habitat. Therefore, no further consultation is required under Section 7 of the Endangered Species Act. Likewise, NRC staff have determined that the proposed action is not the type of activity that has potential to cause effects on historic properties. Therefore, no further consultation is required under Section 106 of the National Historic Preservation Act."

If comments are received from the State or agency, the licensing PM should summarize the comments in the EA. Minor comments could be characterized as "general agreement" or "no objection" by the State or agency. For more extensive comments the licensing PM should summarize the details of the comments and the responses in the EA or place them in a separate document and reference them in the EA. The EA and comment response document should be placed in the NRC Public Document Room to ensure public access.

For rulemaking actions, the draft EA is sent to the State Liaison Officer for comment while the proposed rule is out for comment. This is accomplished through the Office of State and Tribal Programs. The rulemaking task leader is referred to NUREG-0053, "Regulation Handbook" (NRC, 2001) for additional information. Also, consultations beyond that with the State are not typically required for rulemaking actions as these are usually considered administrative in nature. However, the rulemaking task leader should consult the procedures provided in Appendix D.

Conclusion

The conclusion for an EA can be a "finding of no significant impact" (FONSI) or the conclusion can be that there are possible significant impacts from the proposed action. When a FONSI can not be reached an EIS must be prepared; see Section 1.3.1 for guidance on transferring the action to EPAB. The licensing PM is not required to complete the EA if it is determined that an EIS is necessary.

The FONSI is a separate legal finding that is published in the *Federal Register*. The conclusion of the EA supports this finding, however, it does not replace the formal finding that is published in the *Federal Register* (i.e., there is no FONSI section heading in an EA).

A FONSI should include the following language:

> "The NRC staff has concluded that the proposed action **[describe how the proposed action complies with appropriate regulations and brief supporting statement describing minimal impacts, e.g., 'Public exposure to radiation will be less than __% of the limits in 10 CFR Part 20.']**.

> The NRC staff have prepared this EA in support of the proposed action to **[amend or grant license number]**. On the basis of this EA, NRC has concluded that there are no significant environmental impacts and the license amendment does not warrant the preparation of an Environmental Impact Statement. Accordingly, it has been determined that a Finding of No Significant Impact is appropriate."

When a FONSI can not be reached, an EIS must be prepared; see Section 1.3.1 for guidance on transferring the action to EPAB. For completing the EA and documenting the FONSI the licensing PM is referred to Section 3.5-3.6. Additionally, please note that all *Federal Register* notices related to materials licensing actions must be reviewed by OGC (NRC, 2002a).

Sources Used

All references used in preparation of the EA should be listed. Generally, the incoming license request, any correspondence used in preparing the EA, any programmatic documents used in reaching a decision (e.g., Generic/Programmatic Environmental Impact Statement, Standard Review Plan, etc.), and any documents prepared by the staff to reach the conclusion should be referenced here. Additionally, it is helpful to provide ADAMS Accession numbers, if applicable, to assist the public in finding relevant documents.

3.4   EA for Complex Licensing Actions

As discussed in Section 3.2, some NMSS licensing actions are more complex and thus may require additional information in the EA to document the decision. Following is a generic outline for an EA for a complex licensing action (with suggested Section format):

- Introduction (Section 1);
- Need for the Proposed Action (Section 1);
- The Proposed Action (Section 1);
- Alternatives to the Proposed Action (Section 2);
- Affected Environment (Section 3);
- Environmental Impacts (Section 4);
- Mitigation Measures (if applicable, Section 4);
- Monitoring (if applicable, Section 4);
- Agencies and Persons Consulted (Section 5);
- Conclusion (Section 6);
- List of Preparers (Section 7); and
- List of References (Section 8 or can be included at the end of each section).

3.4.1 Introduction of the Environmental Assessment

The introduction should include a brief description of the proposed action, how and when the proposed action was submitted (e.g., license application, license amendment), and by whom the proposed action was submitted. Any unique terms and phrases (e.g., cask, sealed-source, restricted release) should be briefly defined as they are presented. A brief discussion of the relevant NRC regulations should be included. Also, other relevant documents (e.g., another agency's EA or a previous NRC review) should be described in relation to the proposed action and should be clearly referenced.

3.4.2 Need for the Proposed Action

This section describes the applicant/licensee's motivation for submitting the application to the NRC. Examples of need could include a benefit provided to the applicant or any other group if the proposed action is granted and descriptions of the detriment that will be experienced by the applicant or any other group without approval of the proposed action. The need should not be described as a justification for the proposed action over the alternatives. Also, the need should not be stated as NRC approval of a license request.

3.4.3 The Proposed Action

The proposed action should be presented with more detail, including the following:

- Identification of planned activities/phases;

- Location of proposed action;

- The duration of the proposed action (not the duration until the next license renewal), including construction and operation or excavation and/or decommissioning activities, as applicable;

- Relevant and brief descriptions of proposed activities expected to result in impacts should be described in enough detail to support the environmental impacts discussion;`

- Mitigation measures incorporated into the proposed action to avoid significant environmental effects; and

- Maps showing location, facilities, etc., as applicable.

The proposed action is what the applicant/licensee is proposing to do or accomplish with the license amendment. The proposed action is not NRC approval of the license/amendment request.

3.4.4 Alternatives to the Proposed Action

As specified in 10 CFR 51.30(a)(ii), alternatives to the proposed action are developed in accordance with Section 102(2)(E) of NEPA. Although NEPA requirements and CEQ guidance generally address alternatives in the context of an EIS, the same information is generally applicable to an EA. Therefore, alternatives should be considered in an EA (i) if there is some identifiable environmental impact from the proposed action and (ii) if the objective of the proposed action can be achieved in one of two or more ways that will have differing impacts on the environment.

For those actions involving a very small impact, it is reasonable to consider a very limited range of alternatives. In fact, in several decisions, the courts have stressed that the range of alternatives an agency must consider in an EA decreases as the environmental impact of the proposed action becomes less and less substantial. However, no court has held that an agency is excused from considering alternatives if the agency has made a FONSI, and, in fact, considering alternatives is independent of the question of environmental impact.

At a minimum, the no-action alternative must be addressed. The no-action alternative is a discussion of the results from a lack of action (i.e., status quo or the existing state). For example, if the proposed action is the clean-up of a site for unrestricted use, then the no-action alternative is to continue to keep the material licensed and on site, without disposal. More specific guidance on alternatives for an EIS is provided in Section 5.2, *Alternatives*.

3.4.5 Affected Environment

The description of the affected environment should provide a framework for the discussion of impacts (Section 3.4.6, *Environmental Impacts*). Environmental conditions currently existing in the area that could be impacted by the proposed action should be described in this section. The geographic area studied should be identified for each resource. Maps or illustrations may help to provide a clear and concise description. More specific guidance on describing the affected environment for an EIS is provided in Section 5.3, *Description of the Affected Environment*.

3.4.6 Environmental Impacts

The goal of this section is to determine whether there are significant impacts (radiological and nonradiological) for the proposed action and each alternative. Impacts can be direct, indirect, cumulative, long-term and short-term. Direct impacts, or effects, are caused by the action and occur at the same time and place. Indirect impacts, or effects, are caused by the action and are later in time or farther removed in distance, but are still reasonably foreseeable. A detailed definition of direct and indirect effects from 40 CFR 1508.8 states that effects include the following areas of impact: ecological; aesthetic; historical; cultural; socioeconomic; and health. Cumulative impacts are discussed in Section 3.4.6.2 of this guidance. A section on radiological dose impacts should always be provided in the EA and includes both direct and indirect radiation dose impacts to humans. Accident analysis is generally discussed in a SER. The impacts are assessed over the expected lifetime of the action and beyond, if necessary. A scientific basis should be provided; however, there are areas that require professional judgement based on the available information. Where information is incomplete or not available, this should be documented in the EA. Figure 3 provides an overview for analyzing impacts in NEPA documents. A more detailed approach for determining impacts is presented in "Environmental Impact Assessment," (Canter, 1996).

Impacts resulting from each alternative should be briefly described. A table showing the impacts may be useful. Although impacts may exist, they may not be significant, and impacts can be beneficial as well as adverse. Also, an impact that is not significant does not equate to "no impact." Typical impacts may include, but are not limited to:

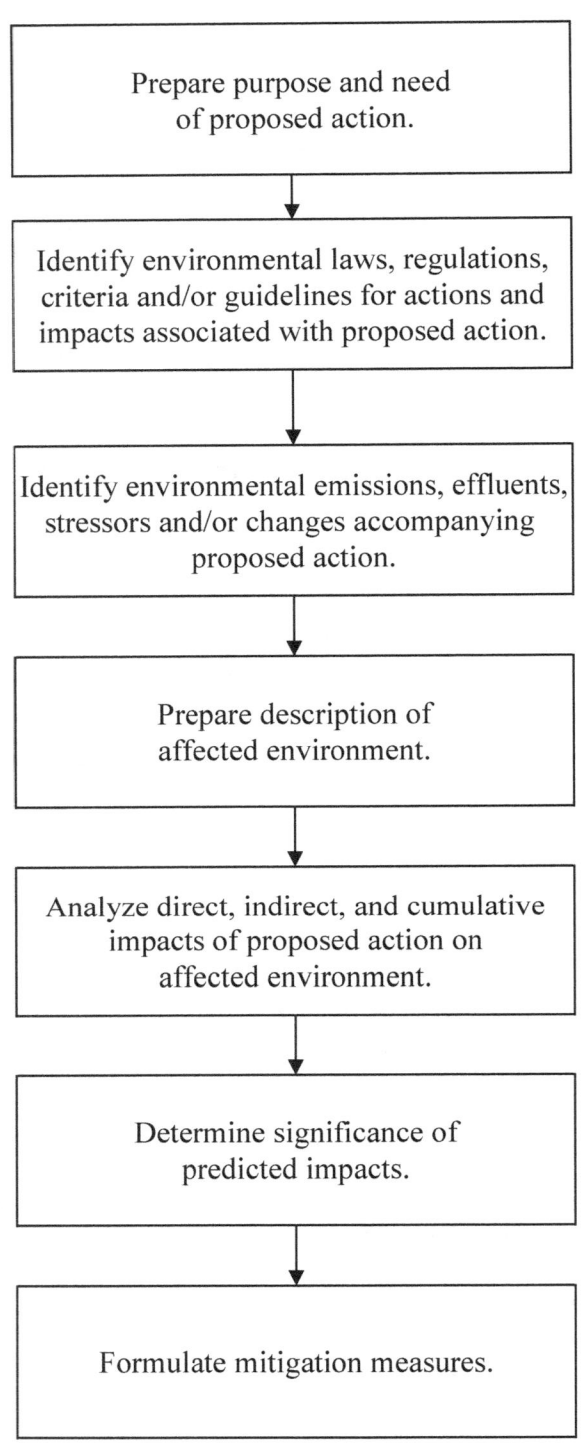

Figure 3: Identifying and analyzing impacts in NEPA documents.

- Increased radiation dose to workers and/or members of the public;
- Degradation of water quality or water supply;
- Habitat destruction;
- Increased air emissions;
- Increased noise;
- Damage or reduced access to cultural resources;
- Changes to local or regional socioeconomic conditions;
- Increased traffic or other transportation effect;
- Increased competition for available resources; or
- Additional population or changing demographics.

For example, describe how the environmental resource (e.g., land or water) is used, how the resource would be affected by the proposed change (e.g., the construction of a building, change in the amount of water taken in by the facility, record keeping or reporting requirements, etc.), and the significance of the relationship between the environmental resource and proposed change. For example, air quality (the environmental resource) would be affected by a release of radioactivity from increased facility effluents and the significance of the release would depend on the types and amounts of the emission. In this example, an appropriate question would be whether the emission for the radionuclide is above the regulatory limits or is a small fraction of the regulatory limits. The section should include an evaluation of radiological and nonradiological impacts. The impacts section should also describe the potential impacts to cultural or historic resources and threatened and endangered species or critical habitat (see Appendix D for more information on consultation requirements). It should clearly state which resources are affected by the proposed action. Likewise, it should clearly state no environmental resources are affected, if that is the case.

If it is determined that a particular action will have no significant environmental impact, then there is no need to consider whether the action will have disproportionately high and adverse impacts on certain populations. Consequently, an environmental justice review will not be completed for an EA where a FONSI is concluded (NRC, 2002b).[6]

For more detailed examples of the types of impacts that should be considered (not necessarily analyzed) see Section 5.4. It is important to understand that all environmental resource areas are not required to be discussed in detail in each EA. The licensing PM should focus the analysis and discussion on resource areas reasonably expected to be impacted.

3.4.6.1 Adverse Impacts

Section 102(2)(C) of NEPA requires consideration of potentially unavoidable adverse environmental impacts should the proposed action be implemented. Discuss both direct and indirect effects and their significance. Potentially adverse impacts of alternatives to the proposed action should also be considered. The discussion of adverse impacts should be thorough, yet brief. Detailed technical information may be incorporated by reference to publicly available materials such as the SER

---

[6]The Commission has directed the staff to develop an environmental justice policy statement. After the policy statement is completed, necessary updates to the environmental justice guidance will be incorporated.

(40 CFR 1502.21). Proprietary data should not be incorporated by reference. It may also be appropriate to discuss potential mitigation measures for adverse impacts (see Section 3.4.7, *Mitigation Measures*).

Both radiological and nonradiological impacts should be discussed. Identify resources that were evaluated but an impact was not found. Impacts may result from construction, operation, and decommissioning phases. Each impacted resource (regardless of significance or size) should be identified, with a rationale provided to explain the determination that the impact(s) are significant or are not significant. The rationale may cite, for example, standards, case history, evaluations or professional judgement. Modeling or other techniques used to predict impacts should be summarized. Impacts are evaluated both onsite and offsite, as well as assessed for cumulative effects. If beneficial impacts are identified, note if a benefit to one party is not viewed as beneficial to a second party.

3.4.6.2 Cumulative Impacts

Unlike an EIS, CEQ regulations do not require an assessment of cumulative impacts in an EA. However, it is suggested that a paragraph be included in the EA that (i) notes the resources with anticipated environmental impacts for the proposed action, (ii) explains that NRC searched for activities that could result in cumulative impacts for those resources, and (iii) states whether there are significant cumulative impacts. A detailed discussion of cumulative impacts in an EIS is in Section 4.2.5.2. A more detailed synopsis is provided in "The NEPA Book," (Bass, Herson, and Bogdan, 2001).

3.4.6.3 Evaluation of Significance

An EA is used to provide sufficient information for determining whether to prepare an EIS or FONSI (10 CFR 51.21, 40 CFR 1501.4) on the proposed action. Impact significance determination involves considering the context and intensity of the impacts. Context means that consideration should be given to what the impacts are, where they will occur, how long they will last, what population or resource is affected, and the carrying capacity of the affected environment. Intensity refers to the impact severity, and can be addressed by a number of criteria delineated in 40 CFR 1508.27. The evaluation of significance should be based on the following considerations:

- Impacts can be both beneficial and adverse. Are there significant adverse impacts despite the existence of beneficial impacts?

- Are there undesirable public health or safety effects?

- Are there unique characteristics of the geographic area such as proximity to historic or cultural resources, park lands, prime farmlands, wetlands, wild/scenic rivers, or ecologically critical areas?

- Are the impacts on the quality of the human environment controversial?

- Are the impacts on the human environment highly uncertain, or do they involve unique or unknown risks?

- Does the proposed action establish a precedent for future actions with significant impacts? Does it represent a decision in principle about a future consideration?

- Is the proposed action related to other actions with individually insignificant, but cumulatively significant impacts? Significance exists if it is reasonable to anticipate a cumulatively significant impact on the environment and cannot be avoided by describing an action as temporary or by breaking it down into small component parts.

- Does the proposed action adversely affect districts, sites, structures, or other objects listed in or eligible for listing in the *National Register*, or will the action result in significant destruction of scientific, cultural, or historical resources?

- Will the proposed action adversely affect an endangered or threatened species or its habitat that has been determined to be critical under the Endangered Species Act?

- Will the proposed action cause a violation of Federal, State, or local law or requirements for the protection of the environment?

If the answer to any of these questions is yes (i.e., impact(s) are significant), then an EIS is normally required. If the answer to all of these questions is no (i.e, no significant impacts are identified), documenting the answers in the EA can be used to prepare the FONSI (see Section 3.6).

The licensing PM, in coordination with management, initially determines whether the proposed action, taking into account reasonable mitigation, will have a significant impact on the quality of the human environment. EPAB is then requested to review the EA, as discussed in Section 3.5. If it is determined that there are no significant impacts, then a FONSI is prepared as discussed in Section 3.6.

If the licensing PM, in coordination with management, determines that the impacts are significant, there are several options for how to proceed. The applicant/licensee may agree in writing to modify the proposed action sufficiently to support a FONSI and a revised EA is prepared. It is possible that the modified proposed action is represented in the alternatives investigated in the initial EA, and only minor changes may be necessary. The applicant/licensee can also agree to mitigate the impacts so that a FONSI can be realized. Mitigation efforts should be clearly identified in the EA document. For example, license conditions and other applicant/licensee commitments may mitigate an impact to permit a "mitigated FONSI". If changes to the proposed action are not available or agreeable to the applicant/licensee to mitigate significant impacts, the licensing PM will forward the project to EPAB who will initiate development of an EIS. The information contained in the EA will form part of the background information for scoping the EIS.

3.4.7 Mitigation Measures

Mitigation measures that could reduce adverse impacts or enhance beneficial impacts should be incorporated in the proposed action to the extent feasible. These mitigation measures may assist in a FONSI. The analysis should address the anticipated effectiveness of these mitigation measures in reducing adverse impacts or enhancing beneficial impacts. The staff should analyze any residual impacts or unavoidable adverse impacts that may remain after mitigation measures have been applied, as well as any further impacts caused by the mitigation measures themselves. Any mitigation measures used to justify FONSIs should be tangible and specific. For example, mitigation measures that avoid, minimize, rectify, reduce over time, or compensate are tangible as opposed to measures that include activities such as further consultation, coordination, and study. A more detailed synopsis is provided in "The NEPA Book," (Bass, Herson, and Bogdan, 2001).

Appropriate monitoring and license requirement for mitigation measures should be identified. Monitoring activities proposed to meet the intent of NEPA should be distinguished from monitoring required by program-specific guidance and/or discretionary monitoring activities.

3.4.8 Monitoring

Any proposed monitoring should be briefly described, including the specific parameters to be monitored (e.g., water quality, noise, species abundance), the frequency (e.g., continuous, once per day), the period of monitoring (e.g., during the entire duration of the site operation), and the action to be taken when thresholds are exceeded. The EA may form the basis for subsequent license conditions on monitoring the proposed action. Monitoring is also discussed for an EIS in Section 5.6, *Environmental Measurements and Monitoring Programs*.

3.4.9 Agencies and Persons Consulted

This section of the EA should list Federal and State agencies and persons consulted. The licensing PM should consult with the affected State just prior to issuing the final EA and should solicit comments on the environmental impact of the proposed action and any other comments the State may have (NRC, 1994). Additionally, the licensing PM is responsible for ensuring that other appropriate agencies are contacted if an action may involve some impact on the natural or physical environment and these consultations should be initiated early in the development of the EA. Appendix D provides a detailed description for the licensing PM to follow for consultations:

- With the appropriate State official;
- Required under Section 7 of the Endangered Species Act; and
- Required under Section 106 of the National Historic Preservation Act.

All consultations should be briefly documented in the EA and should contain (i) the name of the agency or person contacted (consulted), (ii) the date and purpose of the consultation, (iii) a brief summary of the views or comments expressed and the staff's resolution, and (iv) references to publicly available documents containing additional information, as applicable.

For licensing actions that do not affect endangered or threatened species or do not have the potential to cause effects on historic properties the following statement should be considered:

> "NRC staff have determined that the proposed action will not affect listed species or critical habitat. Therefore, no further consultation is required under Section 7 of the Endangered Species Act. Likewise, NRC staff have determined that the proposed action is not the type of activity that has potential to cause effects on historic properties. Therefore, no further consultation is required under Section 106 of the National Historic Preservation Act."

If comments are received from the State or agency, the licensing PM should summarize the comments in the EA. Minor comments could be characterized as "general agreement" or "no objection" by the State or agency. More extensive comments may require the licensing PM to summarize the details of the issues and the resolution of the comments in the EA or to place them in a separate document and reference them in the EA. If there is a differing view that can not be resolved, then it should be noted in the EA. Resolution of the comments should be placed in the NRC Public Document Room to ensure public access.

For rulemaking actions, the draft EA is sent to the State Liaison Officer for comment while the proposed rule is out for comment. This is accomplished through the Office of State and Tribal Programs. The rulemaking task leader is referred to NUREG-0053, "Regulation Handbook" (NRC, 2001) for additional information. Also, consultations will not typically be required for rulemaking actions as these are usually considered administrative in nature. However, the rulemaking task leader should consult the procedures provided in Appendix D.

3.4.10 Conclusion

The conclusion for an EA can be a "finding of no significant impact" (FONSI) or the conclusion can be that there are possible significant impacts from the proposed action. When a FONSI can not be reached an EIS must be prepared; see Section 1.3.1 for guidance on transferring the action to EPAB. The licensing PM is not required to complete the EA if it is determined that an EIS is necessary.

The FONSI is a separate legal finding that is published in the *Federal Register*. The conclusion of the EA supports this finding, however, it does not replace the formal finding that is published in the *Federal Register* (i.e., there is no FONSI section heading in an EA).

A FONSI should include the following language:

> "The NRC staff has concluded that the proposed action **[describe how the proposed action complies with appropriate regulations and brief supporting statement describing minimal impacts, e.g., 'Public exposure to radiation will be less than __% of the limits in 10 CFR Part 20.']**.
>
> The NRC staff have prepared this EA in support of the proposed action to **[amend or grant license number]**. On the basis of this EA, NRC has concluded that there are no significant environmental impacts and the license amendment does not warrant the preparation of an Environmental Impact Statement. Accordingly, it has been determined that a Finding of No Significant Impact is appropriate."

For completing the EA and documenting the FONSI the licensing PM is referred to Section 3.5-3.6. Additionally, please note that all *Federal Register* notices related to materials licensing actions must be reviewed by OGC (NRC, 2002a).

3.4.11 List of Preparers

Identify the principal individuals responsible for the assessment, their professional titles, and the resources they evaluated. For example, "G. Smith, Project Manager in the Division of Waste Management, Health Physics."

3.4.12 List of References

All references (i.e., sources used) used in the preparation of the EA should be listed, including those cited in the text of the EA and those that were not specifically cited but served as useful guidance during document development. NRC guidance, NUREG-0650, "Preparing NUREG-Series Publications," (NRC, 1999) is available and should be useful for determining reference format. Additionally, it is helpful to provide ADAMS Accession numbers, if applicable, to assist the public in finding relevant documents.

3.4.13 Supplemental Information to Environmental Assessment Document

As appropriate, appendices can be included at the end of the EA that include information that is supportive of the findings in the EA. Publicly available information such as letters documenting consultations can be incorporated by reference in the EA.

3.5   Review of a Draft Environmental Assessment Document

The EPAB reviews all NMSS EAs as a final draft document prior to consulting with the State. Section 1.3.1 discusses the process for requesting EPAB review. As discussed in Section 3.4.9, the licensing PM should consult with the affected State before the final EA is prepared. The licensing PM is referred to Appendix D for a suggested procedure to follow in consulting with the State [NOTE: The EA should have a "Conclusion" section with the appropriate language as described in either Section 3.3. or 3.4.10]. After the State has been consulted, the EA is finalized with text noting that the State was consulted along with a summary of the State's comments. If substantive changes are made to the EA as a result of the State or other agency comments, EPAB should review the changes.

In certain circumstances, a draft EA and FONSI may be prepared. Circumstances include those where a FONSI appears warranted for the proposed action but the proposed action is similar to one which normally requires and EIS, the proposed action is without precedent, or the appropriate NRC staff director determines preparation of a draft FONSI will further the purposes of NEPA [10 CFR 51.33(b)]. The draft FONSI should be clearly marked "draft" and should be published in the *Federal Register* and distributed as described in 10 CFR 51.74(a). The *Federal Register* notice must include a request for comments and specify where the comments should be submitted and when the comment period ends (10 CFR 51.119(a)).

3.6   Documenting a Finding of No Significant Impact

If the "Conclusion" of the EA is that no significant impacts exist, a FONSI must be prepared (10 CFR 51.31). The FONSI is not part of the EA, rather the FONSI is a formal finding at the completion of an EA that is published in the *Federal Register* that either includes the EA or a summary of the EA. The FONSI must (10 CFR 51.32): (i) identify the proposed action, (ii) state that the NRC has determined not to prepare an EIS for the proposed action, (iii) briefly present the reasons why the proposed action will not have a significant effect on the quality of the human environment, (iv) include the EA or a summary of the EA, (v) note any other related environmental documents (e.g., application letters, supporting information, consultation letters, etc.), (vi) state that the finding and any related environmental documents are available for public inspection and where the documents may be inspected.

For simple licensing actions (see Section 3.3.), the FONSI is published in the *Federal Register* and may include the complete EA. For complex licensing actions (see Section 3.4), the FONSI is also published in the *Federal Register*, however, a summary of the EA is usually published in lieu of the entire EA.

For rulemaking actions, the FONSI is published in the *Federal Register* for the final rule. The rulemaking task leader is referred to NUREG-0053, "Regulation Handbook" (NRC, 2001) for additional examples.

As mentioned in Section 1.3.1, an initial *Federal Register* notice may be published to announce that NRC received the license application or amendment with an opportunity for hearing. This initial *Federal Register* notice should be referenced when publishing the draft or final FONSI. The FONSI should not include a hearing opportunity notice. The hearing opportunity notice, if required, should be provided in the initial *Federal Register* notice which describes the receipt the of the license application/amendment. Questions on hearing opportunity notices should be directed to the OGC.

As required by 10 CFR 51.35, the final FONSI will be published in the *Federal Register* prior to authorizing the proposed action. In addition the final FONSI must be distributed in accordance with 10 CFR 51.119. The EA and FONSI should be placed in ADAMS and made publicly available.

3.7   References

Bass, R.E.; Herson, A.I.; and Bogdan, K.M.; 2001. "The NEPA Book: A Step-by-Step Guide on how to Comply with the National Environmental Policy Act." Solano Press Books, Point Arena, CA.

Canter, L.W., 1996. "Environmental Impact Assessment." Irwin/McGraw Hill, Boston, MA.

CEQ (Council on Environmental Quality) 1981, "Forty Most Asked Questions Concerning CEQ's National Environmental Policy Act Regulations." CEQ, Executive Office of the President, Washington, D.C. <http://ceq.eh.doe.gov/nepa/regs/40/40P1.HTM >. (December 18, 2002).

NRC (U.S. Nuclear Regulatory Commission), 1994. "State Consultation on Environmental Assessments." Memorandum from Taylor to Russell et. al. U.S. Nuclear Regulatory Commission, Washington, DC. December 6.

NRC, 1997. "Final Generic Environmental Impact Statement in Support of Rulemaking on Radiological Criteria for License Termination of NRC-Licensed Facilities." NUREG-1496. U.S. Nuclear Regulatory Commission, Washingtion, DC.

NRC, 1999. "Preparing NUREG-Series Publications." NUREG–0650. U.S. Nuclear Regulatory Commission, Washington, DC.

NRC, 2001. "Regulations Handbook." NUREG/BR–0053, Revision 5. U.S. Nuclear Regulatory Commission, Washington, D.C. March.

NRC, 2002a. "*Federal Register* Notices For Materials Licensing Actions." Memorandum from Travers to Meserve. U.S. Nuclear Regulatory Commission, Washington, D.C. December 10.

NRC, 2002b. "Staff Requirements (Supplemental) - Affirmation Session, 9:55 A.M., Thursday, November 21, 2002, Commissioners' Conference Room, One White Flint North, Rockville, Maryland, (Open to Public Attendance)." SRM-02-0179. Memorandum from Vietti-Cook to Travers. U.S. Nuclear Regulatory Commission, Washington, DC. December 3.

4 PREPARING AN ENVIRONMENTAL IMPACT STATEMENT: PROCESS

An EIS must be prepared for proposed actions that:

- Are major Federal actions significantly affecting the quality of the human environment (10 CFR 51.20(a)(1));

- The NRC, as a matter of its discretion, has determined that an EIS should be prepared (10 CFR 51.20(a)(2)); or

- Are of the type listed in 10 CFR 51.20 (b).

An EIS provides decision makers and the public with a detailed and objective evaluation of significant environmental impacts, both beneficial and adverse, likely to result from a proposed action and reasonable alternatives. In contrast to the brief analysis in an EA, the EIS includes a more detailed interdisciplinary review. The EIS provides sufficient evidence and analysis of impacts to support the final NRC action in the Record of Decision (ROD; Section 4.10). The draft and final EIS and ROD are made available to the public. Figure 4 outlines the EIS process.

For major licensing actions, as part of the NRC environmental review process, an applicant/licensee should submit information necessary for the environmental review (i.e., prepare an ER, supplement an existing ER, or provide the necessary information with the license application, as appropriate). The environmental PM will review this information and use it to form the basis for assessing environmental impacts of the proposed action and alternatives. Chapters 4 and 5 of this guidance document discuss the EIS process and preparation of the EIS document. Applicants/licensees may find the information in Chapter 6 useful when preparing environmental reports or supplemental environmental reports in support of the proposed action (10 CFR 51.45, 51.60, 51.61, 51.62, 51.66).

For rulemaking actions, there is no applicant to provide environmental information, though in some cases there may be a petitioner for rulemaking who would supply environmental information. Generally, the environmental information needed to support the rulemaking EIS is developed by NRC staff and contractors. Rulemaking EISs usually do not contain site-specific information though generic sites or situations may be described, hence the term Generic Environmental Impact Statement (GEIS). As discussed in Section 1.6.2, "Tiering," rulemaking GEISs should provide ample information regarding bounding conditions and assumptions to allow future reference and tiering

4.1 Project Planning

4.1.1 EIS Team

As stated in Section 1.2.2, EPAB is assigned the responsibility for preparing NMSS EISs. EPAB will designate an EIS or environmental PM who will form an EIS team. The EIS team should include the licensing PM, relevant technical staff who will either prepare or review the EIS, and staff of the Office of Public Affairs and OGC. Also, the environmental and licensing PMs' Section Chiefs, and Licensing

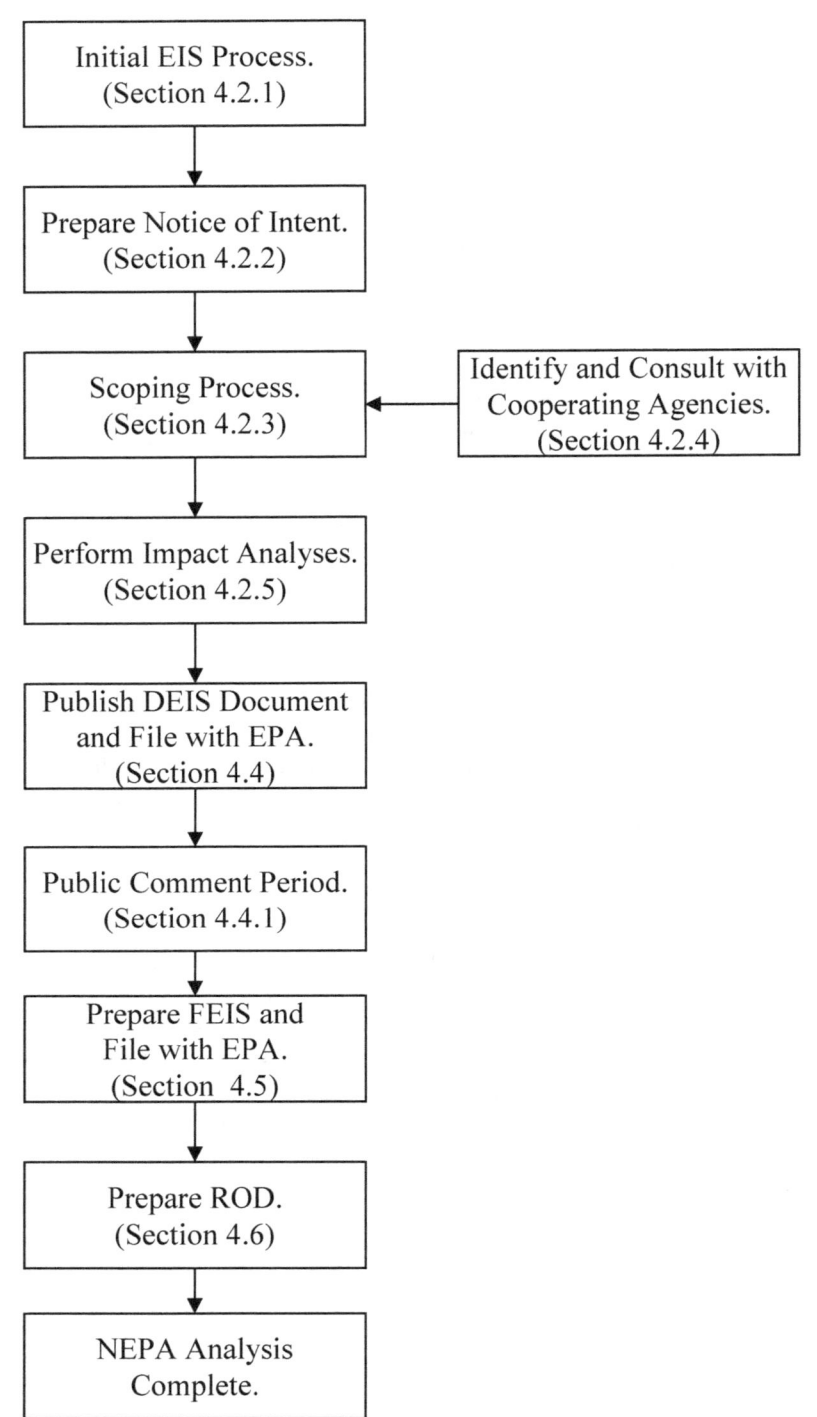

Figure 4: Major steps in the EIS process.

Assistants, and representatives of the Office of State and Tribal Programs (OSTP), and the applicable Regional Office may be part of the team.

The environmental PM, with assistance from the EIS team, should:

- Determine the preliminary scope of the EIS including:

    - developing a purpose and need statement;
    - identifying a list of preliminary alternatives; and
    - developing a list of potentially significant environmental issues.

- Prepare a project plan for the EIS process, including a preliminary schedule for preparing the EIS.

- Assess the need for and provide a recommendation on contractor support.

- Conduct planning for the scoping process to determine:

    - the number and type of scoping meetings;
    - the locations of scoping meetings; and
    - agencies, groups, and individuals to be invited to participate.

- Identify potential cooperating agencies.

- Prepare the notice of intent to be published in the *Federal Register*.

4.1.2 Project Plan

The environmental PM, with assistance from the EIS team, should prepare a project plan for the EIS process. This plan should be used as a basis for managing the project and should be periodically reviewed and modified as needed as the project proceeds. A Gantt chart describing the plan should also be prepared. The plan should include:

- Project purpose and background;

- A description of the principal project tasks and sub-tasks (e.g., planning, scoping, contract acquisition, public participation, technical analyses, preparation of DEIS, etc.);

- Schedule corresponding to the tasks and sub-tasks;

- Resources in staff hours and contract support funds (preferably at the task level);

- Project organization, technical disciplines needed, and responsibilities, including responsibilities for concurrence/approval at each phase; and

- References.

4.1.3 Contractor Support

Because of the complex nature of an EIS and the need for representation, on an EIS team, of several scientific disciplines not normally present amongst NRC staff, the NRC typically uses contractors to assist with preparation of EISs. In some cases, the EIS may be prepared principally by NMSS staff with contractors assisting staff in developing specific portions of the EIS, or a contractor may prepare most of the EIS with the oversight of the environmental PM. Therefore, the EIS team must determine the extent to which contractor support will be required. If the team finds that NMSS staff are not available or do not possess the appropriate expertise, the staff should recommend using an outside contractor to assist in the development of those portions of the EIS for which staff does not have expertise or resources. It is the environmental PM's responsibility to contact the NMSS Program Management, Policy Development and Analysis Staff to discuss the need for contractor support with the appropriate Technical Assistance Program Manager. To best plan and have EIS contractor support in place at the time the license amendment/application is received, the licensing PM should coordinate with EPAB prior to the receipt of the amendment/application.

For rulemaking actions, obtaining contractor support usually begins after Commission approval of the rulemaking plan, though various administrative tasks such as developing the statement of work and independent government cost estimate should be initiated prior to Commission approval.

4.2   EIS Development

4.2.1 Initial EIS Development

Following the acceptance review, as discussed in Section 1.3.3, the environmental PM and other necessary technical reviewers or contractors should begin development of a preliminary draft of the EIS. This effort assists with identification of missing and unclear information, facilitates the preparation of requests to the applicant/licensee for additional information (RAI), streamlines the EIS development, and may assist during the scoping process. Typically, the preliminary draft EIS and the RAIs will not be completed until after the scoping process is complete.

In evaluating the applicant's environmental information, the environmental PM and other technical reviewers should identify and evaluate the quality assurance measures taken by the applicant in collecting and analyzing data. Quality assurance measures, including verification and validation, are also evaluated where computer models have been used to predict environmental consequences of the proposed actions.

Related generic and site-specific EISs should be reviewed to determine if there is a potential for using existing analyses (Section 1.6, *Utilizing Existing Environmental Analyses*). Attention should be given to the bounding conditions (both environmental and nonenvironmental) and related assumptions of these previous analyses to determine if they apply to the new proposed action. This comparison and determination should be briefly described in the EIS. Applicable portions of existing EAs and/or EISs should be incorporated by reference to shorten the length of the EIS.

The identification of potential cooperating agencies should also be made at this time in order to allow full participation in the development of the EIS. A more complete discussion of the role of cooperating agencies is provided in Section 4.2.4, *Consultations and Cooperating Agencies*.

4.2.2 Notice of Intent

After the environmental information and application are accepted for detailed review, the environmental PM will publish the notice of intent (10 CFR 51.26) in the *Federal Register*. The notice of intent is required (10 CFR 51.27) to: (i) state that an environmental impact statement will be prepared; (ii) describe the proposed action and alternatives (if possible); (iii) state whether an environmental report has been filed, and if so, where it is available; (iv) describe the scoping process including the role of participants, whether scoping comments will be accepted, the last date for submitting comments, whether scoping meeting(s) will be held, including the time and place; and (v) state the contact information for the environmental PM. The notice of intent will also briefly describe the proposed action and possible alternatives, describe the proposed scoping process, and state the name and address of the environmental PM. An example is provided in Appendix E.

4.2.3 Scoping Process

Scoping occurs early in the EIS process and provides a means by which the scope of issues to be addressed related to the proposed action are identified. CEQ requirements for scoping are found at 40 CFR 1501.7 and NRC requirements for scoping are found at 10 CFR 51.26-29. Objectives of the scoping process (10 CFR 51.29) include:

- Defining the scope of the proposed action that is to be the subject of the EIS;

- Determining the scope of the EIS and identifying alternatives and significant issues to be analyzed in depth;

- Identifying, and eliminating from detailed study, issues that are peripheral or are not significant;

- Identifying any EAs and other EISs that are being or will be prepared that are related to the EIS under consideration;

- Identifying other environmental review and consultation requirements related to the proposed action;

- Indicating the relationship between the timing of the environmental analyses and the NRC's tentative planning and decision making schedule;

- Identifying any additional cooperating agencies and, as appropriate, allocating assignments for preparation and schedules for completion of the EIS to the NRC and any cooperating agencies; and

- Describing the means by which the EIS will be prepared, including any contractor assistance to be used.

Potential participants in the scoping process are described in 10 CFR 51.28 and typically include:

- The applicant or petitioner for rulemaking (if applicable) in the case of an EIS prepared in support of a rulemaking action;

- Any person who has petitioned for leave to intervene, been admitted as a party, or requested to participate in the proceeding;

- Any Federal agency which has jurisdiction by law or special expertise;

- Affected State and local agencies;

- Affected Federally recognized American Indian Tribes; and

- Any other interested person.

The environmental PM shall ensure that adequate and timely notice of scoping meetings is provided to all potentially interested parties. One of the most frequent complaints about scoping meetings is that participants were not given sufficient notice or did not hear about the meetings until the last minute. In addition to publishing the notice of intent in the *Federal Register*, the meetings should be announced on the NRC's website, in local or regional newspapers, posters around the meeting location, and/or on local radio and television stations at least one week before the meeting is to be held. The environmental PM should consult with the NRC Office of Public Affairs for assistance with newspaper, radio, or television announcements or other avenues for public outreach.

Additional efforts to inform potentially affected groups, such as American Indian tribes and minority and low-income populations, should be undertaken by requesting assistance from tribal leaders, church and community leaders, or other appropriate individuals to disseminate the information. Where such groups may be affected or have expressed concerns, allowing additional time to inform the public before the scoping meeting should be considered. For example, announcements can be included in newsletters read by these groups.

Scoping that is done before an EIS is initiated (e.g., to support an EA preparation) cannot substitute for the formal scoping process after publication of the notice of intent, unless an earlier notice stated clearly that this possibility was under consideration, and the earlier notice expressly provides that written comments on the scope of alternatives and impacts would still be considered. There are no time requirements for the scoping process (10 CFR 51.29 and 40 CFR 1501.7), however; 45 days from the notice of intent should be considered as a minimum length for scoping and accepting scoping comments. If scoping meetings are held, they should be scheduled to ensure that there is a sufficient comment period following the scoping meetings. Comments received after the scoping period has expired should be considered to the extent practicable but may not be able to be included in the scoping report that is issued listing the comments received. For supplemental EISs, scoping is not required (10 CFR 51.92); however, circumstances may indicate that scoping is appropriate (e.g., substantive new or significant information or circumstances) .

4.2.3.1 Scoping Meetings

Although public scoping meetings are not required by NRC's or CEQ's regulations; it is encouraged. NRC practice is to usually hold one or more scoping meetings in the vicinity of the site(s) affected by the proposed action. In certain circumstances (e.g., limited public interest) public scoping meetings may not be held, however, public scoping comments must still be solicited. For rulemaking actions, the scoping meetings should be centrally located to facilitate stakeholder participation. The environmental PM and the EIS team, as appropriate, should visit the site prior to the scoping meeting if they have not already done so in the past. The purpose of a site visit is to familiarize the environmental PM and the team of technical experts who will be preparing the EIS with the site and locale. The environmental PM may visit relevant Federal, State, and local agencies, especially potential cooperating agencies, to obtain information needed to prepare the EIS and to facilitate communication with agencies having an interest in the proposed action. The environmental PM is responsible for coordinating meetings with the licensee and other parties.

The number of scoping meetings to be held should be determined by the types of concerns that have been identified, the areal extent of the proposed action (including direct and indirect impacts), and the amount of controversy associated with the proposed action. For example, if public interest appears to be associated primarily with activities at the site of the proposed action, it may be sufficient to hold a single scoping meeting at a location close to the site. On the other hand, if concerns are raised about transportation of radioactive materials to/from the site, or about other issues having regional or broader impacts, then scheduling scoping meetings in other locales where potential impacts have been identified may be appropriate.

There are no prescribed guidelines for conducting scoping meetings. Development of a format for the meeting should be given careful consideration by the environmental PM and planning team. In preparing for public scoping meetings, PMs should be aware of NRC's "Enhancing Public Participation in NRC Meetings; Policy Statement" (67 FR 36920; NRC, 2002a). Additional guidance is available for conducting public meetings in NUREG/BR-0224, "Guidelines for Conducting Public Meetings" (NRC, 1996a) and NUREG/BR-0297, "NRC Public Meetings" (NRC, 2002b). Relevant guidance is also contained in NRC Management Directive 3.4 "Release of Information to the Public" (NRC, 1999) and NRC Management Directive 3.5 "Public Attendance at Certain Meetings Involving NRC Staff" (NRC, 1996b). Planning for the conduct of the scoping meeting should focus on:

- Goals of scoping;

- Procedures to be used for the meeting;

- Need to focus the discussion in the scoping meeting on:
    - Receiving comments relevant to the proposed activity;
    - Significant issues;
    - Alternatives to be considered;
    - Receiving additional information that participants in the scoping process can provide;
    - Other appropriate concerns;

- Ensuring that the meeting does not become a debate on either the applicant/licensee's justification for the proposed action or the past issues or actions; and

- Use of the EIS in making a decision on the proposed action.

In planning scoping meetings, the environmental PM, with the assistance of the Licensing Assistant, should consider the following to enhance communications:

- Preparing handouts that explain the roles of NRC, cooperating agencies, scoping participants, objectives of scoping, how the meeting is to be conducted, and some background on the proposed action [These handouts can be based on information in the notice of intent, but it should be written in plain language to facilitate communication with a broad audience];.

- Determining the type of meeting format, logistics and setup of the meeting room, procedures for speakers (e.g., registration, order of speaking, time allowed for each speaker), use of handouts, use of public feedback forms, and use of a facilitator;

- Holding an earlier separate meeting with local media reporters to discuss the proposed action, the NEPA process, and the goals of the scoping meeting [Additional guidance is provided in NUREG/BR-0202, "Guidelines for Interviews with the Media" (NRC, 2000).];

- Conducting a poster session (i.e., open house) prior to the scoping meeting to provide an opportunity for one-on-one discussions with interested parties [Ensure that the public understands when comments are being formally transcribed and/or taken.];

- Having the meeting transcribed to document public comments and support the preparation of the scoping report;

- Starting a mailing list for those interested in receiving information about the scoping report, DEIS, etc.; and

- Setting up an EIS project email address to accept comments and a website to house key work products.

Possible formats for conducting scoping meetings include, but are not limited to, the following:

- Facilitated format in which the facilitator opens the meeting with an introduction about the purpose of the meeting and a brief discussion of the background of the proposed action, solicits questions and comments from the audience, guides and focuses the discussion on relevant issues and points, and summarizes the discussion at the end of the meeting;

- Panel format in which a panel of individuals responsible for the EIS and a moderator (often the senior decision maker) introduce the meeting and project similar to the preceding format, but with the panel addressing specific background information on NRC, the project and the decision-making process, and the moderator guiding the meeting (i.e., solicits questions and comments from the audience, guides and focuses the discussion on relevant issues and points, and summarizes the discussion at the end of the meeting); and

- Open house format in which the meeting is set up as a series of discussion stations to address specific issue areas or resources of concern (e.g., public health, ecological resources, socioeconomic) [Attendees should be encouraged to discuss their concerns with appropriate EIS team experts and/or to write down their concerns and turn them in at the meeting. This format can include a formal introduction explaining the purpose of the meeting and directing the attendees to specific areas of interest. It should also include an opportunity for attendees to present oral comments to the NRC and the meeting audience, usually at the end of the meeting.].

4.2.3.2 Scoping Report

In addition to the oral comments gathered at scoping meetings, participants in the scoping process are provided an opportunity to submit written comments on the scope of the EIS. The scoping comment period should extend approximately 30 days after the scoping meeting is held if possible. After the scoping meeting and receipt of written comments, the environmental PM and team will prepare a scoping summary report [10 CFR 51.29(b)]. This report should be a concise summary of the determinations and conclusions reached and should include the following:

- Brief discussion of how the scoping process was conducted, including the dates, locations, and attendance at meetings;

- Discussion of the significant issues and concerns raised;

- Discussion of the alternatives to be evaluated;

- Preliminary schedule for preparing the EIS; and

- Identification of cooperating agencies who will participate in the preparation of the EIS and their roles in EIS preparation.

The environmental PM should send a copy of the final scoping report to each participant in the scoping process. In addition, the report should be included in the EIS as an appendix. The scoping process ends when the issues and alternatives to be addressed in the EIS have been clearly identified and summarized in the scoping report. However, the issues and alternatives can be revised any time before publication of the DEIS.

4.2.4 Consultations and Cooperating Agencies

4.2.4.1 Consultations

Early consultations are essential to: (i) maintaining the planned schedule for completion of the EIS, (ii) gathering complete information, and (iii) identifying potentially significant impacts. Some agencies require 30 days or more to respond to consultation requests and may require additional information from NRC (e.g., photographs, maps, specialized surveys). Consultations may include a number of agencies (e.g., local, county, State, tribal, Federal) which will have information relevant to the site. At a minimum, the following consultations are typically required:

- Section 106 consultation with the SHPO/THPO, Federally recognized American Indian Tribes, or Native Hawaiian organizations for actions with the potential to cause/have effects on historic properties; and

- Section 7 consultation with the FWS for actions which may affect listed species or designated critical habitat.

The environmental PM should document consultations and other sources of information with a brief summary providing the following information: (i) the name of the person, position, and agency consulted; (ii) the date and purpose of the consultation; (iii) a brief summary of the discussion and the staff's resolution; and (iv) references to publicly available documents containing additional information. Consultation letters should be included in an appendix to the EIS. The discussion of the consultation in the EIS should describe why the staff initiated the consultation and summarize the details of the issues and the resolution of the comments in the EIS. The PM is referred to Section 1.4 for a summary of consultation requirements under Section 106 of the National Historic Preservation Act and Section 7 of the Endangered Species Act. Appendix D provides detailed instructions for completing these consultations.

4.2.4.1.1 Interactions with the State

As required by 10 CFR 51.70(c), the staff will cooperate fully with State agencies to reduce duplication between NEPA and State and local requirements. Lists of State Liaison Officers can be found on the OSTP WWW at <www.hsrd.ornl.gov/nrc/asframe.htm>. Often, the State Liaison Officer for NRC is the head of the State agency responsible for radiation protection. Other State contacts (e.g., representatives from the State department of health or environmental quality) who are typically copied on correspondence regarding a license should also be notified of the action.

The environmental PM should contact the NRC Regional Offices to inform them of State interactions. The NRC Regional State Liaison Officers and Regional State Agreement Officers can be found on the OSTP WWW at <www.hsrd.ornl.gov/nrc/contacts/ospstaff.htm>.

OSTP should also be made aware of State interactions. Consulting with the NRC State Liaison Officer is recommended during any consultation with the State. The NRC State Liaison Officer may offer insight to recent NRC-State interactions. During significant interactions with the State, the appropriate NRC State Liaison Officer should receive copies of correspondence with the State.

4.2.4.1.2 Other Consultations

The environmental PM should consult with other agencies that may be impacted or directly involved and identify Federal and State laws that may apply to the site (Section 5.1.4, *Applicable Regulatory Requirements, Permits, and Regional Consultations*). The staff should consult with the agencies responsible for implementing these laws. Examples include sites located on or near Federally controlled land (e.g., Bureau of Land Management), those that affect jurisdictional wetlands (e.g., U.S. Army Corps of Engineers), in proximity to or upstream from National Parks, in proximity to coastal areas subject to the Coastal Zone Management Act, and/or designated as Resource Conservation and Recovery Act (RCRA) or Comprehensive Environmental Response, Compensation, and Liability Act sites by the EPA. If there is a need to contact the EPA, the EPA liaison in DWM should be informed of the contact and the outcome or status. Consultations with American Indian tribes should be conducted in a sensitive manner recognizing the unique government to government relationship that exists based on Federal law and treaties and should be coordinated with the OSTP.

4.2.4.2 Cooperating Agencies

NEPA implementing regulations encourage agencies to become cooperating agencies [10 CFR 51.14(a) and 40 CFR 1501.6, 40 CFR1508.5]. Cooperating agencies can be Federal, State, or local agencies, or an American Indian tribe, if the action may affect a reservation. Frequently, other Federal and/or State agencies have jurisdiction over some aspect of the proposed action. In other cases, an agency may have special expertise in relation to specific environmental issues of concern, and its involvement as a cooperating agency will facilitate the exchange of information and help ensure that applicable requirements are met.

The environmental PM, in consultation with the licensing PM, identifies potential cooperating agencies and requests the participation of agencies at the earliest possible time. Cooperating with Federal, State, and local agencies will reduce duplication between Federal, NRC, and comparable State and local requirements. For potential cooperating agencies that are unfamiliar with nuclear project it may be beneficial for both the licensing and environmental PMs and OGC to meet with representatives of these agencies to explain NRC's mission and other topics relevant to the proposed action.

Contact potential cooperating agencies by letter to determine their interest in participating in the EIS process. Once an agency expresses an interest in becoming a cooperating agency, an agreement should be formalized between NRC and the agency (e.g., a letter of consent, procedural agreement, or a memorandum of understanding) on the cooperating agency's role (e.g., providing information, early review of draft EIS analyses, preparation of EIS sections). It should also be noted that cooperating agencies may have different business practices than NRC and these difference should be addressed as early as possible (e.g., different comment periods for the DEIS).

4.2.4.3 Potentially Interested or Affected Groups

Potentially interested or affected groups, including civic, American Indian tribes, ethnic, special interest groups, and local residents may have special concerns about the proposed action. Identifying those groups and understanding their interests are effective tools for emphasizing important environmental issues and de-emphasizing less important issues. The NRC encourages enhanced public participation in agency decisions.

4.2.5 Impact Assessment

Impacts are assessed for the proposed action and each alternative for each resource described in the affected environment. Consider direct, indirect, cumulative, long-term, short-term, beneficial and negative impacts. To the extent possible, the analysis of impacts should be quantified. Where there is incomplete or unavailable information for evaluating reasonably foreseeable significant adverse impacts, follow the procedures in 40 CFR 1502.22. If an impact can not be quantified it should be described qualitatively. Beneficial impacts may also be identified but both positions should be discussed if a benefit to one party is not viewed as benefit to a second party. A scientific basis should be provided; however, it is recognized that there are areas that require professional judgement based on the available information. A more detailed approach for determining impacts is presented in "Environmental Impact Assessment," (Canter, 1996).

4.2.5.1 Direct and Indirect Impacts

Direct impacts, or effects, are caused by the action and occur at the same time and place. Indirect impacts, or effects, are caused by the action and are later in time or farther removed in distance, but are still reasonably foreseeable. A detailed definition is provided in 40 CFR 1508.8 and describes the following areas of impact: ecological; aesthetic; historical; cultural; economic; social; and health. Both radiological and nonradiological impacts should be discussed. A section on radiological dose impacts should always be provided, including both direct and indirect radiation dose impacts to humans and environmental pathways.

Both geographic and temporal boundaries for each resource should be identified to assist with the discussion of cumulative impact analysis findings discussed below. The EIS author should focus on resource areas where there are impacts. The impacts should be assessed over the expected lifetime of the action (e.g., expected duration of the site) and beyond. Although impacts may exist, they may not be significant. Also, an impact which is not significant does not equate to "no impact." Describe the assessment of impacts from all resources, even those for which an impact was not found.

4.2.5.2 Cumulative Impacts

Cumulative impact is defined as "the impact on the environment which results from the incremental impact of the action when added to other past, present, and reasonable foreseeable future actions regardless of what agency (Federal or non-Federal) or person undertakes such other actions. Cumulative impacts can result from individually minor but collectively significant actions taking place over a period of time" (40 CFR 1508.7).

Examples of cumulative impacts that may be considered:

- Pollutant discharges into surface water;

- Deterioration of recreational uses from loading water bodies with discharges of sediment, nutrients, or thermal effluents;

- Reduction or contamination of ground water supplies; or

- Physically segmenting a community through incremental development.

To determine cumulative impacts, the environmental PM should follow CEQ guidelines as outlined in "Considering Cumulative Effects Under the National Environmental Policy Act" (CEQ, 1997). Other sources of guidance are available from EPA (1999) and the Canadian Environmental Protection Agency (1999).

In general, a cumulative impacts assessment includes the following:

- Determining which resources are affected by the proposed action;

- Identifying other past, proposed, and reasonably foreseeable future actions that either have or might affect those resources;

- Consulting with Federal, State, regional, and local regulators and affected American Indian tribes;

- Identifying likely important cumulative effects;

- Describing cause and effect relationships between stresses (e.g., construction or operation of the facility) and resources;

- Identifying and evaluating potential impacts, but focusing on the most important cumulative impact issues; and

- Determining the magnitude and significance of the proposed action in the context of the cumulative impacts of other past, present and future actions.

If the cumulative impacts are significant, consider avoiding, minimizing, mitigating, or monitoring to address uncertainties.

The following information should be included in the EIS:

- Identification of relevant past, present and reasonably foreseeable future actions, in addition to the proposed action;

- Description of important cause-and-effect pathways;

- Description of significant cumulative impacts and a quantitative description of the magnitude of these impacts;

- Justification for determining that other likely cumulative impacts are not significant;

- For significant cumulative impacts, a discussion of applicant commitments or staff recommendations for actions to minimize environmental harm;

- For significant cumulative impacts, the need for monitoring to reduce uncertainties; and

- Evaluation of reasonable alternatives for cumulative impacts.

4.2.5.3 Evaluation of Significance

Impact significance determination involves considering the context and intensity of the impacts. Context means that consideration should be given to what the impacts are, where they will occur, how long they will last, who is affected, and the carrying capacity of the affected environment. The evaluation of significance should be based on the following considerations (40 CFR 1508.27):

- Impacts can be both beneficial and adverse. Are there significant adverse impacts despite the existence of beneficial impacts?

- Are there undesirable public health or safety impacts?

- Does the proposed action comply with laws, regulations, or executive orders related to historic or cultural resources, park lands, prime farmlands, wetlands, wild/scenic rivers, or ecologically critical areas?

- Are the impacts on the quality of the human environment likely to be controversial?

- Are the impacts on the human environment highly uncertain, or do they involve unique or unknown risks?

- Does the proposed action establish a precedent for future actions with significant impacts? Does it represent a decision in principle about a future consideration?

- Is the proposed action related to other actions with individually insignificant, but cumulatively significant impacts?

- Does the proposed action adversely affect districts, sites, structures, or other objects listed in or eligible for listing in the *National Register* or will the action result in significant destruction of scientific, cultural, or historical resources?

- Will the proposed action adversely affect an endangered or threatened species or its habitat that has been determined to be critical under the Endangered Species Act?

- Will the proposed action cause a violation of Federal, State, or local law or requirements for the protection of the environment?

The environmental and licensing PMs in coordination with management initially determine what impacts the proposed action, taking into account reasonable mitigation, will have on the quality of the human environment. Impact predictions should include comparisons to threshold levels (carrying capacity, maximum concentration limits, etc.). Similar actions, regulations, professional judgement, and public opinion or controversy may all contribute to the evaluation of the significance of the impacts related to the proposed action.

A standard of significance has been established by NRC (see NUREG-1437) for assessing environmental impacts. With the standards of the Council on Environmental Quality's regulations as a basis, each impact should be assigned one of the following three significance levels:

- *Small*: The environmental effects are not detectable or are so minor that they will neither destabilize nor noticeably alter any important attribute of the resource.

- *Moderate*: The environmental effects are sufficient to alter noticeably, but not to destabilize, important attributes of the resource.

- *Large*: The environmental effects are clearly noticeable and are sufficient to destabilize important attributes of the resource.

4.2.6 Request for Additional Information

As discussed in Section 4.2.1, the environmental PM or NRC contractor should develop a preliminary draft of the EIS to assist with the preparation of RAIs. When using a contractor, the outline and draft of the alternatives chapter (Section 5.2, *Alternatives*) should be approved by NRC staff before the contractor begins development of the preliminary draft EIS. The scoping process should be completed before the preliminary draft EIS and the RAIs.

RAI is a term applied to additional information (clarifications and questions) requested of the applicant/licensee in order to complete the environmental and safety reviews. The NMSS goal is to focus the content of RAIs to that additional information necessary to support a regulatory decision. Preparation of a preliminary draft EIS ensures that the necessary information is being requested. RAIs should be documented in a letter to the applicant/licensee. Responses to RAIs should also be in writing.

4.2.7 Format and Content of EIS

NRC's standard format for an EIS is described in Appendix A of 10 CFR 51. Program-specific guidance may identify additional format and content requirements or options. The text of the EIS (not including appendices) should normally be less than 150 pages and for proposals of unusual scope or complexity less than 300 pages. CEQ guidance is provided in 40 CFR 1502.10–1502.18 and 1502.25. An acceptable method of meeting these requirements is provided in Chapter 5, *Preparing an Environmental Impact Statement: Format and Technical Content*.

4.3 Internal Review of the Environmental Impact Statement

Preliminary and final DEIS documents are reviewed by the environmental and licensing PMs, their Section Chiefs, the EIS team, Branch Chiefs, and Division Directors. The Office Director and/or Deputy Office Director may review certain NEPA documents (e.g., EISs involving a great deal of public interest). OGC will review all EIS documents to make a determination of "no legal objection" prior to release to the public. The environmental PM will coordinate the review. The NMSS Division Director (normally the DWM Director) responsible for preparing the EIS is the decision maker for the preliminary recommendation in the DEIS.

After internal review, the initial draft document will be forwarded to the cooperating agencies for review. The document should clearly indicate the following statements on each page: "DRAFT" and "Release of this information to the public or other interested parties is only to be made upon the express permission of the U.S. Nuclear Regulatory Commission." It may be beneficial to meet with cooperating agencies to discuss the preliminary EIS.

The environmental PM and team will revise the DEIS in response to the cooperating agencies comments. A courtesy final DEIS document may be provided to the State and cooperating agencies before the notice of availability is filed with EPA (Section 4.6, *EPA Review*). Reviewers should avoid inadvertent public releases of draft documents.

A preliminary recommendation on the proposed action should be included in the DEIS [10 CFR 51.71(e)]. The preliminary recommendation should be based on the information discussed in 10 CFR

51.71 (e.g., scope of review, analysis of major points of view, status of compliance, and analysis of the environmental effects of the proposed action and reasonable alternatives). In lieu of a recommendation the staff may indicate that two or more alternatives remain under consideration.

4.4   Publishing the DEIS

When submitting the DEIS to the printer provide a copy of the distribution list, as described in NUREG/BR-0188, "Distribution List Descriptions for NRC Reports and Documents," for the initial distribution of DEIS. Sufficient copies must be printed and available for distribution to those who request a copy, either during the scoping process or during the DEIS review period. Copies should also be available for public review in the public electronic reading room. Documents incorporated by reference in the DEIS must also be available for public review in the NRC public document room.

The following NRC standard forms may assist the environmental PM in completing the DEIS.

- Form 335 - Bibliographic Data Sheets;
- Form 426 - Authorization to Publish a NUREG; and
- Form 460 - Request for Graphic Services.

4.4.1   Notice of Availability and Distribution of DEIS

The NRC must publish a *Federal Register* notice announcing the availability of the DEIS as described in 10 CFR 51.117. There are no format or content requirements for a notice of availability other than those associated with the preparation of notices for publication in the *Federal Register* (OFR, 1998). In addition to announcing the availability of the DEIS, the notice of availability must request comments on the proposed action and the DEIS, specify where comments should be submitted, specify when the comment period ends, and when applicable, indicate the dates and location of public meetings to discuss the DEIS. Public comments can be received by mail, email, and on the NRC website, in addition to public meetings. The NRC notice of availability should be coordinated with filing the DEIS with EPA. An example notice of availability is provided in Appendix E.

Beyond the minimum required period of 45 days (10 CFR 51.73), the time period for public comment on a DEIS will be determined based on the potential environmental impact, the extent of the proposed action, any associated controversy, and external time requirements (e.g., statutory deadlines). The environmental PM should ensure the NRC notice of availability and comment period is consistent with EPA's notice of availability (i.e., publication dates in *Federal Register*).

For rulemaking actions, the DEIS is usually issued for a 75-day public comment period to coincide with the comment period on the proposed ruel.

Following completion of the final DEIS, the lead agency is expected to distribute the DEIS for comment to any interested parties. The DEIS will be distributed in accordance with the provisions of 10 CFR 51.74 which include requirements for distribution, news releases, and the notice of availability.

4.4.2 Filing the DEIS with EPA

The DEIS is filed with the EPA's Office of Federal Activities (OFA) who will also publish a *Federal Register* notice of availability. Five copies of the DEIS (including appendices) and a transmittal letter identifying the name and telephone number of the environmental PM should be addressed to:

> US Environmental Protection Agency
> Office of Federal Activities
> EIS Filing Section
> Mail Code 2252-A, Room 7241
> Ariel Rios Building (South Oval Lobby)
> 1200 Pennsylvania Avenue, NW
> Washington, DC 20460

More information on the EPA process is provided at EPA's OFA WWW at <http://www.epa.gov/compliance/nepa/submiteis/index.html> and <http://www.epa.gov/compliance/resources/policies/nepa/fileguide.html>. EPA's review is described in Section 4.6. As described in the EPA filing guidelines (EPA, 1989), the environmental PM should complete distribution prior to transmittal of the DEIS to EPA for filing and review. The EPA notice of availability is a list of all EISs filed the previous week, and is published on Fridays. The NRC notice of availability is a more detailed description of the DEIS and must provide the information in 10 CFR 51.73 and 51.117.

4.4.3 DEIS Public Meetings

Following the publication of the DEIS for public comment, the EIS team usually conducts a public meeting or meetings near the site of the proposed action to receive public comments. In certain circumstances public meetings need not be held (e.g., limited public interest). However, public comments must still be solicited. For rulemaking actions, meetings should be centrally located to facilitate stakeholder participation. The purpose of the public meeting is to allow the staff to explain the contents of the DEIS as well as accept public comments. For actions, such as rulemaking, that may have a national impact, it may be appropriate to schedule and hold a series of public meetings at a number of different locations. For more information see Section 1.7, *Public Meetings*. The following should be considered in preparing for and conducting meetings to gather public comments:

- Scheduling meetings—Provide the public with a reasonable opportunity to review the DEIS prior to the meeting. Generally, the meeting should be held at least 30 days after the EPA notice of filing. However, meetings should not be held so late in the comment period as to preclude attendees from submitting written comments after the meeting.

- Announcing meetings—Announce the dates, times, and locations in the *Federal Register* notice of availability for the DEIS, in a press release to local media, in newspaper advertisements, on NRC's website, and by other means that may be recommended by local officials or groups. Planning for the meeting(s) should be completed before distributing the DEIS.

- Conducting meetings—Records of public meetings should be maintained, including a transcript, a list of attendees (as well as addresses of attendees desiring to be added to the mailing list) and a meeting summary.

- Location of meetings—Hold public meetings at a neutral location (e.g., school auditorium, hotel meeting room, community center, etc.) large enough to handle the expected attendees.

- Format—The format of public meetings will vary. The environmental PM and the EIS team should be prepared to give a summary of the proposed action and potential impacts, allowing time for questions prior to gathering comments from the public.

- Cost—In budgeting for these meetings, the costs should include renting facilities and the necessary equipment, hiring staff (e.g., court reporters, security), and other expenses such as advertisements in the local media.

- The number of people expected to attend the proposed meeting—The number of attendees should be considered when selecting the facility. Guidance is provided in Management Directive 3.5, *Public Attendance at Certain Meetings Involving the NRC Staff* (NRC, 1996b).

- Identify the members of the EIS team who will attend the meeting, and determine their role—For some meeting formats, formal presentations and/or a question and answer session may be appropriate.

- A facilitator—A facilitator may be useful to establish ground rules for conducting the meeting and keeping the meeting focused on the action and DEIS under review. This is especially important for contentious or controversial (local or national) issues.

The following NRC standard forms will assist the environmental PM in preparing for public meetings:

- Form 30 - Request for Administrative Services;
- Form 420 - Request for Premium Cost Mail Service;
- Form 587 - Request for Court Reporting Services; and
- Form 659 - NRC Public Meeting Feedback.

4.4.4 EPA Review

The Clean Air Act (42 USC 7401 et seq.)authorizes the EPA to review proposed actions by Federal agencies in accordance with NEPA and to make those reviews public. Section 309 of the Clean Air Act states that the Administrator [of the EPA] shall review and comment, in writing, on the environmental impact of any matter relating to duties and responsibilities granted pursuant to the Act or other provisions of the authority of the Administrator contained in: (i) legislation proposed by any Federal department or agency; (ii) newly authorized Federal projects for construction and any major Federal agency action (other than a project for construction) to which NEPA applies; and (iii) proposed regulations published by any department or agency of the Federal government. Written comments will be made public at the conclusion of the review. If the EPA Administrator determines that any such legislation, action, or regulation is unsatisfactory from the standpoint of public health welfare or environmental quality, they will publish their determination and the matter will be referred to the CEQ.

If the proposing or "lead" agency does not make sufficient revisions in response to EPA's review of the proposed action and the project remains "environmentally unsatisfactory," EPA may refer the matter to the CEQ for mediation.

The EPA Administrator has delegated responsibility for these reviews to EPA's OFA and the ten EPA Regional Administrators. OFA has developed the following criteria in rating the environmental impacts of a proposed action:

- **LO** - Lack of Objection;

- **EC** - Environmental Concerns - Impacts identified that should be avoided. Mitigation measures may be required.

- **EO** - Environmental Objections - Significant impacts identified. Corrective measures may require substantial changes to the proposed action or consideration of another alternative, including any that was either previously unaddressed or eliminated from the study, or the no-action alternative. Reasons include:

    - violation of a Federal environmental standard;
    - violation of the Federal agency's own environmental standard;
    - violation of an EPA policy declaration;
    - potential for significant environmental degradation; or
    - precedent-setting for future actions that collectively could result in significant environmental impacts.

- **EU** - Environmentally Unsatisfactory - Impacts identified are so severe that the action must not proceed as proposed. If these deficiencies are not corrected in the FEIS, EPA may refer the EIS to CEQ. Reasons include:

    - substantial violation of a Federal environmental standard;
    - severity, duration, or geographical extent of impacts that warrant special attention; or
    - national importance, due to threat to national environmental resources or policies;

EPA uses the following criteria to rate the adequacy of the EIS:

- 1 - Adequate: No further information is required for review;

- 2 - Insufficient Information: Either more information is needed for review or other alternatives should be evaluated. The identified additional information or analysis should be included in the FEIS; or

- 3 - Inadequate: Seriously lacking information or analysis to address potentially significant environmental impacts. The draft EIS does not meet NEPA and or Section 309 requirements. If not revised, or supplemented, and provided again as a DEIS for public comment, EPA may refer the EIS to CEQ.

Additional information on the Section 309 process can be found at EPA's OFA WWW at
<http://www.epa.gov/Compliance/nepa/comments/ratings.html> .

4.4.5 Responses to Comments on the DEIS

Depending on the extent of the proposed action, the anticipated impacts, and the degree of public controversy, the number of written and oral comments received can vary. Comments may lead to modification of the proposed action or alternatives, additional impact analyses, or factual corrections. The FEIS will include responses to individual or grouped substantive comments (10 CFR 51.91).

Comments can be grouped into categories to facilitate responses. All comments must be analyzed, appropriate responses prepared, and the EIS revised as appropriate. Detailed responses should be made to comments that (i) are substantive, (ii) relate to inadequacies or inaccuracies in the analysis or methodologies used, (iii) identify new impacts or recommend reasonable new alternatives or mitigation measures, or (iv) involve substantive disagreements on interpretations of significance. Several typical types of comments and appropriate responses are discussed below.

- Comments on Inaccuracies and Discrepancies—Factual corrections should be made to the DEIS in response to comments that identify inaccuracies or discrepancies in factual information, data, or analyses.

- Comments on the Adequacy of the Analysis—Comments that express a professional disagreement with the conclusions of the analysis or assert that the analysis is inadequate may or may not lead to changes in the FEIS. Public comments may necessitate a reevaluation of analytical conclusions. If, after reevaluation, the environmental PM believes a change is not warranted, the response should provide the rationale for that conclusion.

- Comments That Identify New Impacts, Alternatives, or Mitigation Measures—If public comments on a DEIS identify impacts, alternatives, or mitigation measures that were not addressed in the draft, the environmental PM should determine if they warrant further consideration. If they do, the EIS team should determine whether the new impacts, new alternatives, or new mitigation measures should be analyzed in either the FEIS, a supplement to the DEIS, or a completely revised and recirculated DEIS. If the environmental PM determines that the new impacts, alternatives, or mitigation measures do not warrant further analysis, the response should provide rationale for that conclusion.

- Disagreements With Significance Determinations—Comments may directly or indirectly question the significance or severity of impacts. A reevaluation of these analyses may be warranted and may lead to changes in the DEIS. If, after reevaluation, the environmental PM does not think that a change is warranted, the response should provide the rationale for that conclusion.

- Expressions of Personal Preferences—Comments that express personal preferences or opinions on the proposal do not require a response, however, they should be summarized in the comment section of the FEIS.

4.5   Finalizing the EIS

As a result of public comments the EIS team may determine that additional information is needed from the applicant/licensee before the DEIS can be finalized. Additional RAIs should be provided to the applicant/licensee in writing with the responses to those requests also documented in a letter to the NRC.

Preliminary and final FEIS documents are reviewed by the environmental and licensing PMs, their Section Chiefs, the EIS team, Branch Chiefs, and Division Directors. The Office Director and/or Deputy Office Director may review certain NEPA documents (e.g., EISs involving a great deal of public interest). OGC will review all EIS documents to make a determination of "no legal objection" prior to release to the public. The environmental PM will coordinate the review.

4.5.1   Publishing the FEIS

When submitting the FEIS to the printer provide a copy of the distribution list, as described in NUREG/BR-0188, "Distribution List Descriptions for NRC Reports and Documents," for the initial distribution of FEIS. Sufficient copies must be printed and available for distribution to those who request a copy. Copies should also be available for public review in the public electronic reading room. Documents incorporated by reference in the FEIS must also be available for public review in the NRC public document room.

4.5.2   Distributing the FEIS

Following completion of the FEIS, the lead agency is expected to distribute the FEIS. The FEIS will be distributed as described in 10 CFR 51.93:

- Distribution to:

    - EPA;
    - applicant or petitioner;
    - any other party to the proceeding and each commentor; and
    - appropriate State, regional, and metropolitan clearing houses;

- News releases; and

- Publishing *Federal Register* notice of availability (10 CFR 51.118).

For rulemaking actions, the notice of availability is published in the *Federal Register* notice with the Final Rule.

4.5.3   Filing the FEIS with EPA

The FEIS is filed with the EPA's Office of Federal Activities (OFA) who will also publish a *Federal Register* notice of availability. Five copies of the FEIS (including appendices) and a transmittal letter identifying the name and telephone number of the environmental PM should be addressed to:

US Environmental Protection Agency
Office of Federal Activities
EIS Filing Section
Mail Code 2252-A, Room 7241
Ariel Rios Building (South Oval Lobby)
1200 Pennsylvania Avenue, NW
Washington, DC 20460

More information on the EPA process is provided at EPA's OFA WWW at
<http://www.epa.gov/compliance/nepa/submiteis/index.html> and
<http://www.epa.gov/compliance/resources/policies/nepa/fileguide.html>. As described in the EPA
filing guidelines (EPA, 1989), the environmental PM should complete distribution prior to transmittal of
the FEIS to EPA. The EPA notice of availability is a list of all EISs filed the previous week, and is
published on Fridays. The NRC notice of availability is a more detailed description of the FEIS and
must provide the information in 10 CFR 51.93 and 51.118.

4.5.4 Abbreviated FEIS

If only minor changes are made in the DEIS in response to comments and the changes are confined to
either factual corrections or explanations of why the comments do not warrant further response then an
abbreviated FEIS may be prepared [10 CFR 51.91(a)(3)]. An abbreviated FEIS contains the substantive
comments received on the DEIS, responses to those comments, and an errata section with modifications
and corrections to the DEIS in response to comments. No rewriting or reprinting of the DEIS is
necessary.

4.5.5 Full Text FEIS

If the changes to the DEIS are major, the full-text of the FEIS should be published. The format of the
FEIS is the same as the DEIS, except that the FEIS includes the substantive comments on the DEIS,
responses to those comments, and changes in or additions to the text of the DEIS. The comments are
usually placed in an appendix. The FEIS may incorporate by reference the appendices of the DEIS, if
there are no changes to the appendices. The availability of a full-text FEIS aids subsequent use of the
document for tiering and supplementing purposes.

4.6   Record of Decision

The FEIS and SER form the basis for the NRC decision to approve or deny the applicant/licensee
request. The environmental PM will prepare a concise public ROD (10 CFR 51.102-103) that states: (i)
what the decision is; (ii) all alternatives considered by the NRC and specifying the alternative(s)
considered to be environmentally preferable; (iii) preferences among alternatives based on relevant
factors; (iv) whether the NRC has taken all practicable measures within its jurisdiction to avoid or
minimize environmental harm from the selected alternative and if not, explain why; and (v) summarize
any license conditions or monitoring programs adopted as mitigation measures, if applicable. The ROD
may be integrated into any other record prepared by the Commission in connection with the proposed
action [10 CFR 51.103(c)]. The ROD may also incorporate by reference material contained in an FEIS.

For NRC, issuance of the license, license amendment, or other authorization within the jurisdiction of the NRC such as decommissioning and license termination typically constitute the ROD.

Until the ROD is issued, no action concerning the applicant/licensee proposal will be taken that could have adverse environmental impacts or limit the choice of reasonable alternatives. If NRC is considering an application from a non-Federal entity and is aware that the applicant is about to take an action within the agency's jurisdiction that would meet either criterion (adverse effect or limiting choices), NRC will promptly notify the applicant to stop the action.

The following suggested format satisfies the ROD content requirements specified in 10 CFR 51.103:

- Introductory Material—A cover sheet includes the following information, or most of this information is included at the top of the first page.

  - Title;

  - Docket number and name of applicant /licensee;

  - Preparing office and office location;

  - Cooperating agencies, if any;

  - Signature and title of the responsible official, and signature and title of concurring officials, if any (signature(s) may appear on the last page of the ROD if a cover sheet is not prepared); and

  - Date of signature of approving and concurring officials (this is the official date of the ROD).

- Summary—A summary is needed only if the ROD exceeds 10 pages. It should be a brief synopsis of the ROD.

- Decision [10 CFR 51.103(a)(1), 40 CFR 1505.2(a)]—A clear and concise description of the decision should be prepared. All important aspects or details of the decision should be identified. There should be no ambiguities regarding the specifics of what is or is not being approved.

- Alternatives Including the Proposed Action [10 CFR 51.103(a)(2), 40 CFR 1505.2(b)]—Identify the alternatives considered by the NRC and specify the alternative or alternatives which were considered to be "environmentally preferable."

- Management Considerations [10 CFR 51.103(a)(3), 40 CFR 1505.2(b)]—This section provides the rationale for the decision. Discuss factors, including national policy considerations, NRC's statutory mission, social, economic, technical, and other pertinent considerations weighed in the decision-making process.

- Mitigation and Monitoring [10 CFR 51.103(a)(4), 40 CFR 1505.2 (c)]—Committed mitigation measures and related monitoring and enforcement activities, if any, for the selected alternative are presented here. State whether the NRC has taken all practicable measures within its jurisdiction to avoid or minimize environmental harm from the alternative selected. Measures to avoid or reduce environmental harm which were not selected should also be identified with a brief explanation of why such measures were not adopted. Mitigation and monitoring that will become part of the agency's authorization should be included as stipulations or license conditions in the ROD (i.e., license or license amendment).

- Public Involvement—Briefly describe efforts to seek public views throughout the NEPA process.

4.7   Implementation and Monitoring

Until the ROD has been signed and for at least 30 days following the publication by the EPA of the *Federal Register* notice stating that the FEIS has been filed with the EPA, no action having either an adverse environmental effect or that would limit the choice of alternatives can be taken (10 CFR 51.100-101). Following approval of the ROD and the satisfaction of all other requirements the NRC may approve the action. The approved action must be in accordance with the decision(s) as documented in the ROD. No substantive changes may be made in the implementation of the decision without reconsideration of NEPA compliance needs.

Monitoring and enforcement activities for mitigation measures are generally specified in the ROD as an element of the decision. Most other monitoring activities, however, will not be specified in the ROD. A monitoring plan is recommended for most actions requiring an EIS and should be developed as soon as possible after approval of the ROD.

4.8   References

Canadian Environmental Assessment Agency, 1999, "Cumulative Effects Assessment Practitioners Guide." Canadian Environmental Assessment Agency, Hull, Quebec, Canada. <http://www.ceaa-acee.gc.ca/0011/0001/0004/index_e.htm>. (December 19, 2002).

Canter, L.W., 1996. "Environmental Impact Assessment." Irwin/McGraw Hill, Boston, MA.

CEQ (Council on Environmental Quality), 1997. "Considering Cumulative Effects Under the National Environmental Policy Act." CEQ, Executive Office of the President, Washington, D.C. <http://ceq.eh.doe.gov/nepa/ccenepa/exec.pdf> (December 19, 2002).

EPA (U.S. Environmental Protection Agency), 1989. "Filing System Guidance for the Implementation of 1506.9 and 1506.10 of the CEQ Regulations Implementing the Procedural Provisions of NEPA." U.S. Environmental Protection Agency, Washington, D.C. *Federal Register*: Volume 54, pp. 9592-9594. March 7. Also available at <http://www.epa.gov/compliance/resources/policies/nepa/fileguide.html> (December 19, 2002).

EPA, 1999. "Consideration of Cumulative Impacts in EPA Review of NEPA Documents." U.S. Environmental Protection Agency, Washington, D.C. May. <http://www.epa.gov/Compliance/resources/policies/nepa/>. (December 19, 2002).

OFR (Office of the Federal Register), 1998. "Federal Register Document Drafting Handbook." OFR, National Archives and Records Administration, Washington, D.C. <http://www.archives.gov/federal_register/publications/document_drafting_resources.html>. (December 19, 2002).

NRC (U.S. Nuclear Regulatory Commission), 1996a. "Guidelines for Conducting Public Meetings." NUREG/BR–0224. U.S. Nuclear Regulatory Commission, Washington, DC. February.

NRC, 1996b. "Public Attendance at Certain Meetings Involving the NRC Staff." Management Directive 3.5. U.S. Nuclear Regulatory Commission, Washington, DC. May 24.

NRC, 1999. "Release of Information to the Public." Management Directive 3.4. U.S. Nuclear Regulatory Commission, Washington, DC. December 1.

NRC, 2000. "Guidelines for Interviews with the News Media." NUREG/BR–0202. U.S. Nuclear Regulatory Commission, Washington, DC. June.

NRC, 2002a. "Enhancing Public Participation in NRC Meetings; Policy Statement." U.S. Nuclear Regulatory Commission, Washington, DC. May 28. <http://www.nrc.gov/public-involve/public-meetings/meeting-faq.html>. (December 19, 2002).

NRC, 2002b. "NRC Public Meetings." NUREG/BR–0297. U.S. Nuclear Regulatory Commission, Washington, DC. August.

PAGE INTENTIONALLY BLANK

5 PREPARING AN ENVIRONMENTAL IMPACT STATEMENT: FORMAT AND TECHNICAL CONTENT

This chapter discusses one method of preparing an acceptable EIS. This chapter generally follows the outline of an EIS as described in 10 CFR 51, Appendix A. This EIS format is generally present in all EISs. The information to be provided by the applicant/licensee is described in Chapter 6, *The Environmental Report: Format and Technical Content*.

The scope of the EIS should be balanced against the credible threat to the environment posed by the proposed action (e.g., facility construction, facility operation, or decommissioning). The EIS should present a detailed and thorough description of each affected resource for the evaluation of potential impacts to the environment. Every resource may not receive the same level of detailed review. This is consistent with one of the goals of NEPA, which is to concentrate on the issues that are significant to the proposed action and its potential environmental impacts.

In addition to the EIS, NRC typically prepares a SER to evaluate the radiological impacts of a proposed action. Although there is some overlap between the content of an SER and an EIS, the intent of the documents is different. Since the documents provide input to each other, they must be developed in parallel. This guidance applies to licensing actions. Additional guidance for the preparation of EISs for rulemaking actions is contained in NUREG/BR-0053, "Regulations Handbook" (NRC, 2001).

The rest of this chapter is written to follow the outline of an EIS. Each of the following section headings describe the types of information usually included in the EIS. It is acceptable to combine chapters to make a more readable document, as long the required information (10CFR 51.70 and 51.71) is present. Following is an example table of contents:

Executive Summary

Chapter 1 Introduction
    1.1   Purpose of and Need for the Proposed Action
    1.2   The Proposed Action
    1.3   Scope of This Environmental Analysis
         1.3.1   Issues Studied in Detail
         1.3.2   Issues Eliminated from Detailed Study
    1.4   Applicable Regulatory Requirements, Permits, and Regional Consultations
    1.5   Comments on the Draft Environment Impact Statement

Chapter 2 Alternatives
    2.1   Process Used to Formulate Alternatives
    2.2   Proposed Action
    2.3   No-Action Alternative
    2.4   Other Reasonable Alternatives
    2.5   Alternatives Considered but Eliminated
    2.6   Comparison of the Predicted Environmental Impacts
    2.7   Preliminary Recommendation

Chapter 3 Description of the Affected Environment

Chapter 4 Environmental Impacts

Chapter 5 Mitigation Measures

Chapter 6 Environmental Measurements and Monitoring Programs
    6.1   Radiological Monitoring
    6.2   Chemical Monitoring
    6.3   Ecological Monitoring

Chapter 7 Cost-Benefit Analysis

Chapter 8 Summary of Environmental Consequences

Chapter 9 List of Preparers

Chapter 10 Distribution List

Chapter 11 List of References

Appendices

5.1   Introduction of the EIS

The following background information should be provided:

- Proposed action and relevant background;

- Explanation of why this action requires an EIS;

- Brief history of the facility (if not a new application) or program, as appropriate; and

- List of the other alternatives considered.

5.1.1 Purpose and Need for the Proposed Action

This section explains why the proposed action is needed. It describes the underlying need for the proposed action and should not be written merely as a justification of the proposed action, nor to alter the choice of alternatives. Another common mistake is to identify compliance with NEPA and CEQ regulations as the need. Examples of need include a benefit provided if the proposed action is granted or descriptions of the detriment that will be experienced without approval of the proposed action. In short, the need describes what will be accomplished as a result of the proposed action.

5.1.2 The Proposed Action

This section should briefly describe the proposed action, including the name of the applicant/licensee, the title of the project, the location (with a map), and the schedule. This section should also describe the desired outcome or goal of the proposal. For example, at a decommissioning site, the licensee must meet the 10 CFR 20, Subpart E, radiological criteria for license termination. For a new fuel cycle facility, the applicant/licensee must meet the 10 CFR 70 criteria.

5.1.3 Scope of This Environmental Analysis

This section describes the scoping process. The scoping process, as described in Section 4.2.3, will result in the scope of the EIS.

The following information should be included in the EIS:

- History of the planning and scoping process for this project;

- Discussion of public concerns;

- List of cooperating agencies and the reasons they became cooperating agencies;

- List of other Federal, State, local, and other organizations contacted; and

- Summary of related EISs, EAs and other relevant documents, such as the SER and includes mention of former EAs for the site and GEISs used in tiering.

5.1.3.1 Issues Studied in Detail

The scoping process identifies two categories of issues - those that need to be studied in detail (but do not necessarily result in significant impacts) and those that can be eliminated from detailed study because the impacts are minimal. Resources (ground water, historic properties, ecological resources, etc.) are generally the same as issues. However, a resource could be split into two issues - for example, short-term socioeconomic impacts due to construction and long-term socioeconomic impacts to land use. To make the EIS less like an encyclopedia and more issue-driven, it is recommended that the environmental analysis be separated into these two categories. This approach leads to an EIS that emphasizes the principal results of the analysis, and these two sections (5.1.4.1 and 5.1.4.2) are a summary of the conclusions regarding environmental impacts.

This section provides a summary of the issues that require more detailed study. Among these issues are those that may result in significant short- or long-term impacts. Each issue and the conclusion regarding its potential impact are described briefly (no more than a few paragraphs). A more detailed analysis of the impacts should be presented in the EIS chapter, "Environmental Impacts."

5.1.3.2 Issues Eliminated from Detailed Study

This section summarizes the issues that were found to have minimal short- and long-term impacts. Each issue and the conclusion regarding its potential impact are described briefly in one or two paragraphs. If necessary, the issues eliminated from detailed study are discussed further in an appendix. The reader is referred to the appropriate EIS section in the appendix if there is further explanation.

5.1.4 Applicable Regulatory Requirements, Permits, and Regional Consultations

The staff review includes identification of applicable consultations, approvals, and authorizations (and the relevant agencies). The review should include: (i) determination of the status of the consultations and/or authorizations; (ii) identification of environmental concerns; and (iii) evaluation of potential administrative problems that could delay or prevent agency authorization.

The staff should:

- Identify all Federal, State and local permits, licenses, approvals, and other entitlements that must be obtained in connection with the proposed action.

- Produce a summary of compliance with applicable environmental quality standards and requirements that have been imposed by Federal, State, and local agencies.

Table 1 illustrates a sample format for summarizing the list of permits, licenses, approvals, entitlements and consultations and their status. The table can be used by the reviewers to identify areas of environmental concern and determine applicant/licensee compliance with existing standards and regulations. In some circumstances (e.g., a potential problem in State siting authorizations), the environmental PM may need to prepare additional information to fully cover the subject material. If it is uncertain whether a Federal permit, license, approval, or other entitlement is necessary, the DEIS will so indicate (10 CFR 51.71(c)).

Table 1. Sample format for Federal, State, and local authorizations and consultations

Agency	Authority	Activity Covered	Status*
US. Army Corps of Engineers	Clean Water Act, Section 404	Dredge and Fill Permit	Approval to be obtained
U. S. Fish and Wildlife Service	Endangered Species Act	Biological Assessment	Undetermined at present
State Historic Preservation Office	National Historic Preservation Act	Consultation	Initial consultation complete
*This field to be filled in based on the consultations with relevant agencies.			

5.1.5 Comments on the Draft Environmental Impact Statement

In the Final Environmental Impact Statement (FEIS), include a summary of the major public comments on the DEIS. Include details on the comments and responses in an appendix.

The following information should be included in the FEIS:

- Date(s):

    - DEIS was submitted to EPA;
    - Notice of intent was published in the *Federal Register*; and
    - DEIS was made available to the public;

- Methods used to publicize the availability of the DEIS;

- Schedule of public meetings held on the DEIS, including location, date, and time; and

- Summary of major comments and responses.

5.2   Alternatives

This section introduces alternatives that could also accomplish the need for the proposed action. This section should discuss the no-action alternative, the proposed action, and the reasonable alternatives. Alternatives should be included that will avoid or minimize adverse effects upon the quality of the human environment.

All alternatives, including the no-action alternative, should receive equal and objective treatment. The phrase "range of alternatives" includes all reasonable alternatives (including the no-action alternative) to the proposed action, as well as those other alternatives that are eliminated from detailed study, with a brief discussion of the reasons for eliminating them. Reasonable alternatives are those alternatives that meet the proposal objectives and applicable environmental standards and are technically feasible.

The number of alternatives considered is generally small (e.g., three to five alternatives). The discussion of alternatives should include similar types of descriptions as for the proposed action. Describing the alternatives in a parallel format for presentation makes the comparisons clear to the reader. The alternatives should also be summarized in a table for efficiency and clarity.

The environmentally preferable alternative is the alternative that offers the best combination of minimized damage to the biological/physical environment and protection of historic, cultural, and natural resources. The environmentally preferred alternative may not necessarily be the same as the proposed action or chosen alternative because of many factors, including cost/benefit analyses, mitigating factors, and legal considerations.

A preliminary recommendation on the proposed action should be included in the DEIS [10 CFR 51.71(e)]. The preliminary recommendation should be based on the information discussed in 10 CFR 51.71 (e.g., scope of review, analysis of major points of view, status of compliance, and analysis of the environmental effects of the proposed action and reasonable alternatives). In lieu of a recommendation the staff may indicate that two or more alternatives remain under consideration.

5.2.1 Process Used to Formulate Alternatives

Briefly describe the process used to formulate alternatives - licensee submittals, public input during the scoping process, interdisciplinary discussions, etc. As a general matter, the staff has broad discretion in consideration of alternatives in the EIS and is not limited to considering only those alternatives proposed by the applicant/licensee. However, the selection of an alternative solely because it is economically superior to the proposed action is inconsistent with past NRC practice. In general, the staff should include all reasonable alternatives to the proposed action with the purpose of identifying those that are environmentally superior (NRC, 1997).

5.2.2 Proposed Action

This section describes the proposed action in greater detail, usually what the applicant/licensee proposes in their license application. It should not include descriptions that are more appropriate in the purpose and need section.

This section should also describe the facility and location. It should provide a detailed description of the facility's geographical location including an overview map of within 50 miles of the site, a more detailed map within 5 miles, and a map of the facility layout. The layout description should identify all buildings and pertinent features. The site features most likely impacted (or to cause impacts) by the proposed action should be described in detail. The location description will establish a geographical point of reference for other resource descriptions (e.g., land and water use, local ecology, or socioeconomic).

The facility descriptions should include the nature and extent of present and proposed operations at the site, facilities that might be constructed, modified, or impacted as a result of the proposed action, summary description of the facility operations (including the types and methods of material movement from one part of the site to another), and identification of the radionuclides and hazardous materials used, including where and how they are stored, handled, utilized, and disposed. A complete description of the facility support systems (e.g., electrical power, gas supply and water supply etc.) should be provided. This section should also describe nonradiological and radiological contamination at the site/facility and provide a discussion of background radiological characteristics. Discuss any accidents that may have occurred during operation and their impacts.

5.2.3 No-Action Alternative

This section describes the no-action alternative along with a description of the major impacts. For the no-action alternative, the proposed action would not take place. This serves as a baseline for comparing alternatives. For example, in a license application proposing new construction and/or activities the no-action alternative would be to not grant the license (i.e., no construction or activity). In a license renewal situation, the no-action alternative would be to deny the amendment request (the licensee would still have to comply with other applicable requirements). For certain decommissioning actions, the no-action alternative (i.e., not perform the decommissioning activity) may not be a reasonable option and detailed analysis of impacts is not usually performed.

5.2.4 Other Reasonable Alternatives

This section describes other reasonable alternatives to the proposed action and a summary of their major impacts. A description of reasonable alternatives depends on the nature of the proposal and the facts in each case. As discussed in 40 CFR 1502.14, the emphasis is on reasonable rather than whether the applicant/licensee likes or is capable of carrying out a particular alternative. Reasonable alternatives include those that are practical or feasible from the technical and economic standpoint and using common sense, rather than simply desirable from the standpoint of the applicant/licensee (CEQ, 1981).

5.2.5 Alternatives Considered but Eliminated

This section summarizes the alternatives that were eliminated from detailed study, with a brief discussion of the reasons for eliminating them. The section does not need to be exhaustive, but should at least discuss alternatives that have been proposed in applicant/licensee documents, public meetings, and related correspondence. If the no-action alternative is not a reasonable option due to legal, safety, or considerations, it should also be discussed in this section.

5.2.6 Comparison of the Predicted Environmental Impacts

This section describes and compares all alternatives. Discussion of the impacts of the alternatives should be limited to a descriptive summary of the impacts to all resources. The information contained in this section should also be incorporated into a summary table.

5.2.7 Preliminary Recommendation

As described in 10 CFR 51.71(e) the DEIS should normally include a preliminary recommendation on the proposed action. This recommendation should be based on the information and analyses contained in the DEIS and reached after consideration of the environmental impacts of the proposed action and reasonable alternatives. In lieu of a recommendation the staff may indicate that two or more alternatives remain under consideration.

5.3 Description of the Affected Environment

The description of the affected environment focuses on baseline conditions, i.e., the status quo. The baseline conditions will be used to assess the impacts discussed in Section 5.4, *Environmental Impacts.*

The following environmental resources should be considered, as appropriate in preparing the EIS:

*   Land use;
*   Transportation;
*   Geology and soils;
*   Water resources;
*   Ecology;
*   Meteorology, climatology, and air quality;
*   Noise;
*   Historical and cultural resources;

- Visual/scenic resources;
- Socioeconomic;
- Public and occupational health; and
- Waste management.

5.3.1 Land Use

This section should describe existing and planned (without the proposed action) land uses for the site and vicinity. The EIS should include maps that provide locations of schools, hospitals, farming areas, and other land uses important to impact assessment. A discussion of possible conflicts between Federal, State, regional, and local (and in the case of a reservation, American Indian tribe) land-use plans, policies, and controls for the site should also be included.

5.3.2 Transportation

If transportation is an important issue, it may be necessary to develop a separate section on transportation instead of incorporating this information in the land use or socioeconomic section. This section should describe transportation resources at and around the facility. The EIS should describe transportation infrastructure as it is important for considering impacts such as site workers commuting and transportation of materials. This section should describe local roads and highways, railroads, navigable rivers, and provide information on current levels of traffic.

5.3.3 Geology and Soils

The section should provide a brief summary of regional and site geology. Reference the SER for additional details. The EIS should discuss regional and local structure, the site stratigraphy, characteristics of the soil, major structural and tectonic features (e.g., faults), any other significant geological conditions, local and regional seismicity data, and volcanism.

5.3.4 Water Resources

This section describes the water resources, including surface and ground water hydrology, water use, and water quality. The EIS should describe the surface water bodies and ground water aquifers that could be affected by the proposed action and should consider both regional and site specific data. The EIS should provide a map showing the relationship of the site to major hydrogeologic systems. Describe flood plains, wetlands, streams, reservoirs, etc. The EIS should include a description of site-specific and regional data on the characteristics of surface and ground water quality in sufficient detail to provide the necessary data for other reviews dealing with water resources. The EIS should include a discussion of water quantity available for use and possible conflicts between Federal, State, regional, local and American Indian tribe, in the case of a reservation, water-use plans, policies, and controls for the site.

Consumptive water uses that could affect the water quality and supply of the proposed action or that may be adversely affected by the proposed action should be identified including water source, locations of diversions and returns, amount used and seasonal use patterns, and water rights. Also, recreational, navigational, and other non-consumptive water uses including those that could be affected by offsite area construction and operation by location, activity, and amount used, and seasonal use patterns should be provided. Finally, this section should identify water uses that provide potential pathways for both radiological and nonradiological effluents including water sources, locations of diversions for consumptive uses, locations of receptors for non-consumptive uses, amount used, and seasonal use patterns.

Additional sources of information should be utilized when needed to complete the analysis. Sources include local water supply companies or agencies, river basin commissions, State agencies (e.g., water resources, fish and wildlife), Federal agencies (e.g., U.S. Army Corps of Engineers and the U.S. Geological Survey) and American Indian tribal agencies. From the information gathered from these resources, compile and tabulate water uses by the categories and characteristics, but limit the analysis to consideration of past, present, and known future water uses. The EIS preparer should ensure that water-use data and information are adequate to serve as a basis for assessing the impacts of proposed project construction and operation on consumptive and non-consumptive water uses.

5.3.5 Ecology

This section describes the principal ecological (terrestrial and aquatic) features of the site and vicinity, transportation corridors, and region, with emphasis on the plant and animal communities that may be affected by the proposed action. This information should include transient and migratory species to reflect any seasonal variations in ecological populations.

The EIS should include a description of ecological resources (e.g., endangered, threatened, and important species including estimates of their abundance) and special habitat needs (e.g., cover, forage, and prey species) of species in the area. The EIS should include information on the species and habitats as described in Table 2.

A complete species list may be prepared as an appendix to the EIS. Additionally, a summary should be provided of the consultations with appropriate Federal, State, regional, local, and American Indian tribal agencies, including the FWS and the State fish and wildlife agency, with details provided in an appendix.

In addition to NEPA, Section 7 of the Endangered Species Act, and 50 CFR 402, require the NRC to meet certain requirements in the protection of endangered and threatened species and critical habitat. The environmental PM is referred to Appendix D for a detailed description for completing the Section 7 consultation requirements.

5.3.6 Meteorology, Climatology, and Air Quality

This section should provide a detailed description of the meteorological/climatological conditions and baseline air quality of the site and region around the proposed action.

The EIS should provide a description of relevant meteorological, climatological, and air quality data sufficient to establish regional and local baseline conditions for the site. The information provided in this section will be used in the analysis of impacts on air quality. The EIS should include:

- Description of the existing regional air quality for completeness and accuracy; and

- Air pollutants for which there are non-attainment or maintenance areas in the region.

Table 2. Important species and habitats

Species	Habitat
Rare species • Listed as threatened or endangered at 50 CFR 17.11 (Fish and Wildlife) or 50 CFR 17.12 (Plants). • Proposed for listing as threatened or endangered, or is a candidate for listing. • Listed as a threatened, endangered, or other species of concern by the State or States in which the proposed facilities are located. Commercially or recreationally valuable species. Species that are essential to the maintenance and survival of species that are rare and commercially or recreationally valuable (as defined previously). Species that are critical to the structure and function of the local terrestrial and aquatic ecosystems. Species that may serve as biological indicators to monitor the effects of the facilities on the terrestrial and aquatic environments.	Wildlife sanctuaries, refuges, or preserves, if they may be adversely affected by the proposed action. Habitats identified by State or Federal agencies as unique, rare, or of priority for protection, if these areas may be adversely affected by the proposed action. Wetlands (Executive Order 11990), flood-plains (Executive Order 11988), or other resources specifically protected by Federal regulations or Executive Orders, or by State regulations. Land areas identified as "critical habitat" for species listed as threatened or endangered by the FWS.

5.3.7 Noise

This section describes the current sources and levels of noise. This discussion should be consistent with the terms concepts described in EPA (1974) and American Society for Testing and Materials (1996) material. The EIS should include a comparison of the estimated sound levels to appropriate limits. The EIS should provide a description of the analysis and assessment of current and historical trends, noise levels, applicable sound level standards, and current practices to minimize adverse noise impacts.

5.3.8 Historic and Cultural Resources

In addition to NEPA, Section 106 of the National Historic Preservation Act, and 36 CFR 800, require the NRC to meet certain requirements in the protection of cultural and historical resources. The environmental PM is referred to Appendix D for a detailed description for completing the Section 106 consultation requirements.

The environmental PM should consider historic, archaeological, and traditional cultural resources in sufficient detail to provide the basis for subsequent analysis and assessment of possible impacts. Historic and cultural resources include districts, sites, buildings, structures or objects of historical, archaeological, architectural, or cultural significance. The environmental PM should be aware of results of any surveys conducted; the location and significance of any properties that are listed in or eligible for inclusion in the *National Register* as a historic place; and any additional information pertaining to the identification and description of historic properties that could be impacted by the proposed action.

The construction, subsequent operation, and/or decommissioning of a facility could impact historic properties directly (e.g., destruction or alteration of the integrity of a property) or indirectly (e.g., prohibiting access or increasing the potential for vandalism). In considering the areal extent of the review, note that a facility can have a visual or audible effect on historic resources that are located some distance from the proposed facility.

The NRC can authorize the applicant/licensee to initiate consultations with the SHPO to determine if there are any historic properties listed in or eligible for inclusion in the *National Register*. The review should also include historic properties included in State or local registers or inventories and any additional important cultural, traditional, or historic properties. If necessary, during scoping, discuss with the SHPO any organizations or individuals that might be able to assist in identifying and locating archaeological and historic resources (for example, university and American Indian tribal archaeological and historical staffs).

If a property appears to meet the *National Register* criteria, or if it is questionable whether the criteria are met, the staff may request, in writing, an opinion from the U.S. Department of the Interior regarding the property's eligibility for inclusion in the *National Register*. The request for determination of eligibility should be sent directly to the Keeper of the *National Register*, National Park Service, US. Department of the Interior, Washington, D.C. 20013-7127. Guidance from the National Park Service can be found on the WWW at <http://www.cr.nps.gov/nr/publications/> (NPS, 2003a).

The Archeology and Ethnography Program of the National Park Service may be a useful source of expertise in the area of historic and cultural preservation and is staffed with professionals who may be able to assist the NRC staff in the environmental review and in analyzing the results of the applicant's surveys and investigations. Further information can be found on the WWW at <http://www.cr.nps.gov/aad/> (NPS, 2003b).

To discourage property vandalism and scavenging, particularly in the case of archaeological sites, it may be necessary to provide information to the SHPO for handling in a confidential manner. Summary information, which does not include site-specific information, could be included in the EIS documentation. State and tribal laws/policies addressing the handling of confidential and sensitive information vary and may not coincide with Federal regulations, regardless of how the information is marked by a licensee/applicant or NRC. Hence, specific requests for maintaining confidential or sensitive information should be discussed with States and tribes.

Contact the Advisory Council on Historic Preservation if guidance is needed, if there are substantial impacts on important properties, in the event of a disagreement, or if there are issues of concern to American Indian tribes or Native Hawaiian organizations.

The EIS should summarize the applicant's and staff's review and include the following information:

- Historic properties listed in or eligible for inclusion in the *National Register*;

- Historic properties included in State or local registers or inventories;

- Any additional important cultural, traditional, or historic properties;

- Efforts to locate and identify previously recorded archaeological and historic sites;

- Overall results and adequacy of any surveys (archival or field) that were conducted by the applicant; and

- A list of organizations and individuals contacted by the applicant/licensee or the staff who provided significant information concerning the location of cultural and historic properties.

5.3.9 Visual/Scenic Resources

This section describes the landscape characteristics, manmade features, and view of the proposed action site.

The EIS should include the staff's assessment of the applicant/licensee's rating of the aesthetic and scenic quality of the site in accordance with the Bureau of Land Management (BLM) Visual Resource Inventory and Evaluation System (BLM, 1984, 1986a, 1986b, 2002). Particular attention should be paid to viewsheds and likely activities in the proposed action that may reduce the visual/scenic resource. This description will be used later in evaluating the impacts of the proposed action and alternatives on visual/scenic resources.

5.3.10 Socioeconomic

This section describes population distribution and community characteristics within the region that are likely to be affected by the proposed action and each alternative. The EIS should include descriptions of relevant past and current population distributions. Both permanent and transient populations should be identified. Describe low-income and minority populations. This description will be used to assess impacts (including radiological impacts) on social, economic, and community resources.

The following information should be presented in the EIS:

- Population characteristics (e.g., ethnic groups, and population density);

- Economic trends and characteristics, including employment and income levels;

- Housing, health and social services, and educational resources;

- Area's tax structure and distribution; and

- Summary of any coordination with appropriate local and regional agencies or groups who collect these types of data.

5.3.11 Public and Occupational Health

This section describes levels of background radiation, major sources and levels of background chemical exposure, occupational injury rates, and health effects studies performed in the region.

The EIS should include information on current background levels, historical exposure levels for actions similar to the proposed action, and a summary of any public health studies performed in the region sufficient to establish baseline information for analysis of impacts to public and worker health.

5.3.12 Waste Management

This section summarizes the historical baseline data regarding the production, handling, packaging, and shipping of waste. The EIS should discuss disposal practices for solid, hazardous, radioactive, and mixed wastes including disposal capacity. The baseline conditions will be used in the analysis of nonradiological and radiological impacts due to waste management.

5.4 Environmental Impacts

This section summarizes the known and potential impacts (e.g., direct, indirect, and cumulative) of the proposed action and each alternative. These impacts should consider normal operational events as well as reasonably foreseeable accidents (e.g., design basis events for 10 CFR 72 licensees or credible consequence events for 10 CFR 70 licensees). When analyzing impacts, resources should be considered separately, and where necessary, in combination (e.g., noise impacts on wildlife, or transportation impacts on land use).

Activities (i.e., construction, operation and decommissioning) should be evaluated in sufficient detail to determine the significance of potential impacts and to recommend how these impacts should be treated in the process (e.g., consideration of alternative designs or practices that would mitigate adverse environmental impacts).

Evaluation of each identified impact should result in one of the following determinations:

- The impact is small and mitigation is not required.

- The impact is adverse but can be mitigated by specific design or procedure modifications that the reviewer has identified and determined to be practical.

- The impact is adverse, cannot be successfully mitigated, and is of such magnitude that it should be avoided.

5.4.1 Land Use Impacts

This section should describe the impacts to land use for each alternative. The following information should be presented in the EIS:

- Long-term restrictions of land use resulting from the proposed action and long-term changes in land use of the site and vicinity;

- Short-term changes in land use of the site and vicinity;

- Restrictions or modifications of lands classified as floodplain, wetlands, or coastal zone; as described in Section 5.4.5;

- Conflicts between Federal, State, local, or American Indian land use plans;

- Mitigation measures for adverse impacts (e.g., earth leveling, revegetation, landscaping, cleanup and disposal of debris, erosion control structures, land management practices, stabilization of spoil piles, and stabilization of dikes on cooling lakes); and

- Prevention of current or planned mineral resources exploitation (e.g., sand and gravel, coal, oil, natural gas, or ores).

5.4.2 Transportation Impacts

This section describes transportation impacts, both incident-free and accidents, for each alternative. The discussion of transportation impacts should include all phases of the project from any newly constructed transportation corridors or increased usage of existing corridors for construction of the project, through transportation issues during operation of the facility, to any increased transportation which may occur during decommissioning. Guidance for this review is provided in NUREG-0170, "Final Environmental Statement on the Transportation of Radioactive Material by Air and Other Modes" (NRC, 1977a).

The analysis should consider transportation mode, routes, risk estimates, and impacts of transportation on the environment, including increases and decreases in usage of transportation corridors. Consider new construction that may be needed to upgrade existing or create new transportation routes and modes.

The following information should be included in the EIS:

- Transportation mode, routes, and risk estimates and impacts and their significance for each alternative;

- Potential mitigative measures proposed to decrease the transportation impacts for each alternative including the degree that these measures are effective in mitigating the impacts for each alternative; and

- Comparison of the offsite dose consequences and resulting health effects as calculated by the applicant/licensee and those contained in the SER. Review of the dose consequence analysis including the direct, indirect, and cumulative socioeconomic impacts and the impacts to biota.

The EIS author should coordinate this section with the transportation analysis conducted for the SER.

5.4.3 Geology and Soils Impacts

This section summarizes potential geological impacts, which may also be assessed in the staff's SER. The analysis should be incorporated by reference from the SER. Examples of geological environmental impacts include soil compaction, soil erosion, subsidence, landslides, and disruption of natural drainage patterns.

5.4.4 Water Impacts

This section describes the surface and ground water impacts from the proposed action and each alternative, including water use, and water quality. The description should include consideration of site-specific and regional data on the water-use characteristics, water quality, and hydrology of ground and surface water. The description should include an analysis and evaluation of construction, operation and decommissioning activities in sufficient detail to determine the significance of potential water impacts and to recommend how these impacts should be treated in the process (e.g., consideration of alternative designs or practices that would mitigate adverse environmental impacts). The details of these supporting analyses (e.g., actual environmental measurements, modeling assumptions and results) should be disclosed by reference or placed in an appendix to the EIS.

The analysis should consider the following:

- Changes to the hydrological system that could cause ground and surface water impacts at and near the site [The analyses of water system alterations and water-supply/water consumption comparisons should be included. These changes could include water quantity and availability, water flow, and movement patterns, and erosion, deposition, and sediment transport. All water system characteristics should be included in this analysis (e.g., all sources of water, points of discharge, and water diversions) that modify the availability of water. The analyses should include short-term and long-term effects and include discussions of flood plain alterations.];

- Impacts resulting in reduced water availability [Identify the location of those water users likely to be affected, and consider adverse effects (e.g., lowered ground water table, reduced well yields, lowered surface-water levels at intake structures) to determine their impacts on individual water users or water-use areas. The reviewer should consider seasonal requirements for water and temporal variations in water availability. The reviewer should also consider the potential for an incompatibility between water availability as affected by project activities and existing and known future water rights and allocations. The nature and extent of these future water inequalities should be identified.]; and

- Water quality potentially impacted by modifications to the ground and surface water system or users [The analysis should consider short-term effects as well as long-term effects caused by each alternative. Alternatives should be identified that avoid adverse effects and incompatible development in the flood plain. The reviewer should identify alternative designs, construction and operational practices, or procedures that could mitigate or avoid the impacts.]

The following information should be included in the EIS:

- A description of the impacts to water quality/availability in the region;

- Direct, indirect, and cumulative impacts from each alternative (radiological and nonradiological);

- Assessments of both short- and long-term effects;

- A comparison of water quality impacts to appropriate standards;

- A description of the aquatic transport and diffusion characteristics relevant to the alternatives which should include references to the models used and identification of the input data considered;

- A dose assessment of the radiological impacts based on sufficient aquatic transport parameters and population data; and

- A description of mitigative measures for water quality/availability impacts.

5.4.5 Ecological Impacts

This section summarizes the ecological (terrestrial and aquatic) impacts of the proposed action and each alternative. An assessment of both onsite and offsite activities including transportation corridors should be provided. The assessment should be in sufficient detail to: (i) predict and evaluate the significance of potential impacts to important species and their habitats; and (ii) evaluate how these impacts should be considered in the process.

The analysis should consider activities that:

- Create obstacles to the movements of vertebrates or result in increased dispersal of invertebrate species known to be important as disease vectors or pests;

- Disturb benthic (i.e., lake, sea, or river bottom) areas [All dredged areas or areas affected by dredging may be considered as temporarily lost habitat, therefore dredging should be limited, if possible.];

- Potentially increase surface run-off [Good construction practices will generally control surface run-off. Where drainage courses represent an especially important resource, attention should be given to measures for their protection.];

- Involve dewatering of wetlands [Guidelines under the Federal Water Pollution Control Act (i.e., Section 404 of the Clean Water Act), the Coastal Zone Management Act of 1972, and the Marine Sanctuaries Act of 1972 should be followed in evaluating the significance of dewatering on wetlands. Generally, dewatering of biologically productive wetlands may be considered an adverse impact that should be avoided. The percentage loss of such wetlands in the region should be considered to place the loss in perspective for the licensing decision. Because of the importance of wetlands, alternatives to avoid any loss of this habitat should always be considered. Contact with the U.S. Army Corps of Engineers, Regulatory Branch District Office, may be necessary to obtain a wetland delineation and/or permit to modify a wetland.];

- Involve dredge spoils and placement of fill [Drainage from dredge spoil areas should comply with existing EPA guidelines. The analysis should consider whether adequate practices have been provided for management of this stage of construction. Filling of biologically productive wetlands should generally be avoided. Dumping of dredge spoils should be performed under the cognizance of the EPA and the Regulatory Branch District Office of the U.S. Army Corps of Engineers.];

The depth and extent of the input to the EIS should be governed by the attributes of the ecological resources that could be affected and by the nature and magnitude of the expected impacts to those resources.

The following information should be included in the EIS:

- Results of consultations performed as required by Section 7 of the Endangered Species Act;

- Loss of habitat for endangered or threatened species in the context of guidelines under the Endangered Species Act of 1973 [Where loss of habitat for commercially or recreationally important species occurs, the environmental PM should consider the effects on the harvestable crop. It should generally be concluded that loss of up to 5 percent of such habitat in the site vicinity will have negligible impact on the crop and need no further analysis. Where losses exceed 5 percent, the environmental PM should consider the loss in relation to regional abundance of these species.];

- Practices to minimize soil erosion and the number of hectares disturbed;

- Clearing of vegetation from stream banks, making certain that it is limited to that necessary for placement of structures or decontamination of hazardous or radiological constituents;

- Secondary impacts on wildlife, such as altered behavior resulting from construction noise, in addition to direct impacts on animals such as loss of habitat and road kills; and

- Lost important terrestrial and aquatic species and habitats from the viewpoints of their uniqueness within the region under consideration, relative impacts, and long-term net effects.

5.4.6 Air Quality Impacts

This section describes the air quality impacts from the proposed action and each alternative and the atmospheric transport and diffusion processes important in determining impacts. The description should include an analysis and evaluation of construction, operation and decommissioning activities in sufficient detail to determine the significance of potential air quality impacts and to recommend how these impacts should be treated in the process (e.g., consideration of alternative designs or practices that would mitigate adverse environmental impacts). The details of this supporting analyses (e.g., actual environmental measurements, modeling assumptions and results) should be disclosed. Adverse cumulative effects of each alternative should be identified.

The analysis should utilize models and assumptions that have been approved or recognized for use in appropriate regulatory guidance for air quality monitoring and/or dose assessments. At least one annual data cycle should be used for transport and diffusion calculations. Data should be presented in the appropriate periods. For example, if emissions are continuous, annual data should be used; if emissions are intermittent, consideration should be given to the frequency and duration of the event. Data, such as averages and extremes, should be based on a period of record that represents long-term conditions in the area.

The following information should be included in the EIS:

- A description of the impacts to air quality in the region;

- Direct, indirect, and cumulative impacts from each alternative (radiological and nonradiological);

- Assessments of both short- and long-term effects (hourly and annually);

- A comparison of air quality impacts to appropriate standards;

- Description of necessary air permits;

- A description of the atmospheric transport and diffusion characteristics in the region and at the site, which should include references to the models used and identification of the input data considered;

- A dose assessment of the radiological impacts based on sufficient meteorological and population data;

- A description of visibility impacts; and

- A description of mitigative measures for air quality impacts.

5.4.7 Noise Impacts

This section describes the analysis and assessment of predicted noise levels from the proposed action and each alternative. The description should include an analysis and evaluation of construction, operation and decommissioning activities in sufficient detail to determine the significance of potential noise impacts and to recommend how these impacts should be treated in the process (e.g., consideration of alternative designs or practices that would mitigate adverse environmental impacts). Details of supporting analyses (e.g., actual environmental measurements, modeling assumptions and results) should be disclosed. Known and/or predicted adverse direct, indirect, and cumulative effects of each alternative should be identified.

If the site is remote from communities (ecological and human) and does not represent an audible intrusion, and it is found that the applicant can comply with appropriate guides and standards, these facts should be stated with only a very brief discussion noting that under these conditions noise impacts will be minimal. If the foregoing conditions are not met, or if there are no applicable standards, predicted impacts should be described along with conclusions regarding the significance of the effect on the community.

If the site is located near communities (ecological and human) and noise impacts are a potential concern, the following information should be included in the EIS:

- A comparison of the current equivalent sound levels in the vicinity of the proposed action and applicable sound level standards (from consultation with Federal, State, regional, local, and affected American Indian tribal agencies) with predicted noise levels (e.g., sound contour maps) reported as L_{eq} or L_{dn} using the dBA scale;

- Major sources of noise (for locations described above), including all models, assumptions and input data;

- Proposed methods to reduce noise levels (as appropriate); and

- Estimated cumulative effects.

5.4.8 Historic and Cultural Impacts

This section describes the staff's assessment of potential impacts of proposed project activities on historic properties and cultural resources in the site and vicinity. Historic properties include districts, sites, buildings, structures, or objects of historical, archaeological, architectural, or traditional cultural significance (NPS, 2002). In addition to NEPA, Section 106 of the National Historic Preservation Act, and 36 CFR 800, require the NRC to meet certain requirements in the protection of cultural and historical resources. The environmental PM is referred to Appendix D for a detailed instructions on completing the Section 106 consultation requirements.. Elements of Section 110 of National Historic Preservation Act require Federal agencies to manage and protect identified, eligible historic properties located on lands under their jurisdiction. A source of expertise in the area of historic and cultural preservation is the Archaeology and Ethnography Program of the National Park Service, Department of Interior (NPS, 2003b).

The environmental PM should consider the following in preparing the analysis:

- Construction and/or operation activities that could result in potential impacts to historical properties or cultural resources;

- Proposed activities to ensure that the applicant is committed to using currently acceptable practices to minimize impacts;

- 36 CFR 800, which describes how to Federal agencies meet the statutory responsibilities under Section 106 of the National Historic Preservation Act;

- That there are generally two types of impacts on a resource: direct impacts (e.g., destruction during excavation), and indirect impacts (e.g., visual impact, denial of access, or increased potential for vandalism);

- Certain properties are not eligible for inclusion in the *National Register*, and assistance from the SHPO/THPO, the Office of Archaeology and Historic Preservation, or other qualified individuals may be necessary to complete the analysis;

- Adequacy of proposed methods to mitigate any adverse impacts on these resources such as alternative locations, designs, practices, or procedures that would mitigate predicted adverse impacts;

- Cost of the recovery required by the Historic and Archaeological Preservation Act of 1974 in the consideration of alternatives;

- Evaluations that may not justify preservation of the resource [In such cases the environmental PM may request that the applicant recover archaeological, historic, architectural, and cultural data related to the resource. This recovery may include recording by photographs and measured drawings, archaeological excavations to uncover data and material, removal of structures or salvage of architectural features, and other steps that will ensure full knowledge of the lost resource. Salvaged artifacts and materials should be deposited where they are of public and educational benefit.];

- Any procedures developed by the applicant/licensee that will be used during construction in the case of discovery of previously unidentified cultural resources;

- The potential for human remains to occur in the project areas should be evaluated [An inadvertent discovery of such items during construction may necessitate a work stoppage and consultation under Native American Graves Protection and Repatriation Act procedures.]; and

- Circumstances in which to contact the Advisory Council on Historic Preservation if guidance is needed (i.e., if there are substantial impacts on important properties, in the event of a disagreement, or if there are issues of concern to American Indian tribes or Native Hawaiian organizations).

The following information should be included in the EIS:

- Results of consultations performed as required by Section 106 of the National Historic Preservation Act;

- If appropriate, a statement that properties listed in or eligible for inclusion in the *National Register* will not be affected;

- Discussion of potential impacts (e.g., direct, indirect, and cumulative) to properties that are listed in or eligible for inclusion in the *National Register*;

- Description of any adverse impacts on historic properties not eligible for inclusion in the *National Register*; and

- Description of any measures and controls that are available to limit adverse impacts.

5.4.9 Visual/Scenic Impacts

This section describes the significant impacts on visual quality resulting from the proposed action and each alternative. Scenic qualities are impacted by surface disturbance, which creates a contrast with the natural environment. The greater the amount of ground disturbance, the greater the impact to scenic quality. The description should include an analysis and evaluation of construction, operation and decommissioning activities in sufficient detail to determine the significance of potential visual/scenic impacts and to recommend how these impacts should be treated in the process (e.g., consideration of alternative designs or practices that would mitigate adverse environmental impacts). The environmental PM may assess the licensee's rating of aesthetic and scenic quality of the site in accordance with the BLM Visual Resource Inventory and Evaluation System (BLM, 1984, 1986a, 1986b, 2002) as appropriate.

The EIS should describe the impacts of the proposed action and each alternative on the visual quality of the vicinity. Significant visual quality impacts should be thoroughly described, while less-significant, yet still noteworthy, impacts can be summarized. The EIS should describe how impacts could be minimized. The description of mitigation measures should provide a short discussion of costs of the mitigation measures.

5.4.10 Socioeconomic Impacts

This section describes the socioeconomic impacts within the region. Based on these descriptions, the environmental PM should identify and analyze project-induced changes to demographic, regional, community, social, political, and economic systems.

The EIS should describe impacts from the proposed action and each alternative relative to the current and predicted population distributions. Both permanent and transient populations should be considered.

The following information should be presented in the EIS:

- Impacts to population characteristics (e.g., ethnic groups, and population density);

- Impacts to economic trends and characteristics, including employment and income levels;

- Impacts to housing, health and social services, and educational resources; and

- Impacts to the area's tax structure and distribution.

5.4.11 Environmental Justice

The Commission has directed the staff to develop an environmental justice (EJ) policy statement. After the policy statement is completed, necessary updates to the EJ guidance will be incorporated. In the interim, the following draft guidance on environmental justice is being provided.

This section evaluates environmental impacts on low-income or minority populations by proposed project activities if disproportionately high low-income or minority populations are identified. Impacts that may have environmental justice implications may include health, ecological (including water quality and water availability), social, cultural, economic and aesthetic resources.

The EIS should follow the detailed guidance provided in Appendix C. The EIS should include a discussion of the methods used to identify and quantify impacts on low-income and minority populations, the location and significance of any environmental impacts during construction on populations that are particularly sensitive, and any additional information pertaining to mitigation. The following information should be included in the EIS:

- An assessment (qualitative or quantitative, as appropriate) of the degree to which each minority or low-income population is disproportionately receiving adverse human health or environmental (including socioeconomic) impacts during construction, operation, or decommissioning as compared with the other population in the vicinity. In addition, there should be an assessment comparing the impacts with the larger overall geographic area encompassing all of the alternative sites.

- An assessment (qualitative or quantitative, as appropriate) of the significance or potential significance of such environmental impacts on each low-income and minority population. Significance is determined by considering the disproportionate exposure, multiple-hazard, and cumulative hazard conditions.

- An assessment of the degree to which each low-income and minority population is disproportionately receiving any benefits compared with the entire geographic area.

- A discussion of any mitigative measures for which credit is being taken to reduce environmental justice concerns.

- When alternative sites are being evaluated, the same reviews should be available for each site.

- A brief description of pathways by which any environmental impact during construction may interact with cultural or economic facts that may result in disproportionate environmental impacts on low-income and minority populations.

5.4.12 Public and Occupational Health Impacts

5.4.12.1 Nonradiological Impacts

This section describes the pathways by which nonradiological releases could be transmitted to the environment and ultimately transferred to living organisms. The analysis should be based on the information from Section 5.3.12, *Public and Occupational Health* to assess the potential impacts, mitigation measures and cumulative effects. The analysis should consider potential pathways for the transfer of nonradioactive materials from the proposed action and alternatives to the environment and ultimately to living organisms. The analysis should identify all pathways necessary to calculate public and occupational exposure.

The following information should be included in the EIS:

- A description of chemical sources (location, type, strength);

- Estimates of public and occupational exposures, a brief discussion of how the estimates were calculated, and a comparison of these exposures with the requirements of 40 CFR 190 and 29 CFR 1900;

- Brief discussion of environmental monitoring programs to verify compliance; and

- Discussion of mitigative measures and cumulative effects and how requirements have been met.

5.4.12.2 Radiological Impacts

This section summarizes the direct and indirect radiological impacts, mitigation measures, and cumulative impacts from each alternative. This section is divided into Sections 5.4.12.2.1, *Pathway Assessment* and 5.4.12.2.2, *Public and Occupational Exposure Impacts*.

5.4.12.2.1 Pathway Assessment

This section should describe the pathways by which radiation and radioactive releases can be transmitted to the environment and ultimately transferred to living organisms. The scope and depth of the review should include consideration of: (i) the pathways by which radioactive releases can be transported to individual receptors; (ii) the location of these receptors; and (iii) the credible threat to the environment posed by the facility, action, or activity.

The following information should be included in the EIS:

- Typical pathways by which radioactive materials could be transported from the various alternatives to receptors in unrestricted areas;

- Pathways identified as important for the various alternatives and a brief discussion of the staff's analysis to determine these pathways;

- Nearest receptors identified by the reviewer; and

- Brief discussion of food production, processing, and consumption in the area.

5.4.12.2.2 Public and Occupational Exposure Impacts

This section should describe the radiation dose to humans. The staff reviewer should evaluate the baseline information (Section 5.3.12, *Public and Occupational Health*) to assess the potential impacts, mitigation measures, and cumulative impacts.

The following information should be included in the EIS:

- Description of radiation sources (location, type, strength) related to the proposed action;

- Estimates of dose to an average member of the critical group and occupational dose estimates, a brief discussion of how the estimates were calculated, a comparison of these doses with the requirements of 10 CFR 20, and the conclusions with respect to compliance with 10 CFR 20;

- Brief discussion of environmental monitoring programs to verify compliance (Section 4.5, *Environmental Measurements and Monitoring Programs*);

- Discussion of mitigative measures; and

- Comparison of the offsite dose consequences and resulting health effects for reasonably foreseeable (i.e., credible) accidents as calculated by the applicant and those contained in the SER. The environmental PM should coordinate this section with the analysis conducted for the SER.

5.4.13 Waste Management Impacts

This section describes the staff's review, analysis, and evaluation of the applicant/licensee's solid, hazardous, and radioactive waste management program including the assessment of impacts resulting from storage or transportation. A discussion of mixed waste is also included in this section.

The EIS should be of sufficient depth and detail to confirm, with reasonable assurance, the quantitative impact of the waste management systems. Facility owners/operators are required by RCRA regulations to maintain sufficient information to identify their mixed wastes. The information required includes RCRA waste codes for the hazardous components, the source of the hazardous constituents, a discussion of how the waste was generated, the generation rate and volumes of mixed waste in storage, and any information used to identify mixed wastes or make determinations that the wastes are prohibited by land disposal restrictions. Each owner/operator is required (under RCRA regulations) to develop a waste-minimization plan that identifies process changes that can be made to reduce or eliminate mixed wastes, methods to minimize the volume of regulated wastes through better segregation of materials, and the substitution of nonhazardous materials.

The following information should be presented in the EIS:

- Descriptions of the sources, types, quantities, and composition of solid, hazardous, radioactive and mixed wastes expected from the proposed action;

- Description of proposed waste management systems designed to collect, store, and dispose of all wastes generated by the proposed action;

- Anticipated disposal plans for all wastes (i.e., transfer to an offsite waste disposal facility, treatment facility, or storage onsite); and

- A waste-minimization plan that identifies process changes that can be made to reduce or eliminate waste. This should contain a description of methods to minimize the volume of waste.

5.5   Mitigation Measures

Mitigation measures that could reduce adverse impacts should be incorporated in the proposed action and alternatives (40 CFR 1502.14(f) and 1508.20). The mitigation measures discussed in the EIS must cover the range of impacts of the proposal. The measures must include such things as design alternatives that would decrease pollution emissions, construction impacts, esthetic intrusion, as well as relocation assistance, possible land use controls that could be enacted, and other possible efforts. Mitigation measures must be considered even for impacts that by themselves would not be considered "significant." If the proposed action as a whole is considered to have significant effects, all of its specific effects on the environment (whether or not "significant") must be considered, and mitigation measures must be developed where it is feasible to do so (CEQ, 1981). Mitigation measures should be tangible and specific. For example, mitigation measures that avoid, minimize, rectify, reduce over time, or compensate are tangible as opposed to measures that include activities such as further consultation, coordination, and study. A more detailed synopsis is provided in "The NEPA Book," (Bass, Herson, and Bogdan, 2001).

All relevant, reasonable mitigation measures that could improve the project should be identified, even if they are outside the jurisdiction of the NRC. The probability of the mitigation measures being implemented and the time line for their implementation should also be discussed for both NRC activities and activities under the jurisdiction of another agency.

The anticipated effectiveness of these mitigation measures in reducing adverse impacts, the technical feasibility, and the cost-benefit of any recommended mitigation measures should be discussed in the EIS (costly actions that would yield only minor environmental benefits should not be recommended).

5.6   Environmental Measurements and Monitoring Programs

This section describes the environmental measurements and monitoring programs for the proposed action. A more detailed description of the monitoring program is usually provided in the SER prepared in parallel with the EIS.

Mitigation monitoring activities proposed to meet the intent of NEPA [40 CFR 1505.2(c)] should be clearly distinguished from monitoring required by program-specific guidance and/or discretionary monitoring activities.

5.6.1 Radiological Monitoring

This section describes the proposed monitoring program utilized to characterize and evaluate the radiological environment, to provide data on measurable levels of radiation and radioactivity, and to provide data on principal pathways of exposure to the public.

The following information should be provided in the EIS:

- Maps or aerial photographs of the facility with proposed monitoring and sampling locations clearly identified along with effluent release points;

- Brief description of the monitoring program including:

    - Number and location of sample collection points, measuring devices used, and pathway sampled or measured;

    - Sample size, sample collection frequency, and sampling duration; and

    - Type and frequency of analysis including lower limits of detection;

- Principal radiological exposure pathways (Section 5.4.12.2.1, *Pathway Assessment*); and

- Location and characteristics of radiation sources and radioactive effluent (liquid and gaseous, from Sections 5.4.4, *Water Impacts* and 5.4.6, *Air Quality Impacts*).

5.6.2 Physiochemical Monitoring

This section should describe the proposed monitoring program to characterize and evaluate the chemical and physical environment, to provide data on measurable levels of chemicals and baselines for physical parameters of importance (i.e., weather conditions).

The purpose of a chemical environmental monitoring program is to provide a basis for evaluating changes in the environment from the proposed action. The baseline monitoring program should characterize the environment before the proposed action so that a reasonable comparison can be made after the proposed action begins. The baseline program can also be used for all or some of the operational chemical environmental monitoring program.

The EIS should describe the applicant's/licensee's chemical monitoring program. Two aspects of monitoring should be considered:

- Baseline monitoring is used to support the applicant's baseline descriptions and provide information for operational comparison; and

- Operational monitoring establishes the impacts of operation of the facility and detects any unexpected impacts arising from facility operation.

Each of these aspects is discussed in greater detail below.

Baseline Monitoring

Information from the applicant's/licensee's baseline monitoring program is used to aid in the assessment of site acceptability/condition and to support the staff's database to identify impacts that could result from the selected alternative. Generally, data are needed on a seasonal basis and should be sufficient to characterize seasonal variations throughout at least one annual cycle.

The environmental PM should analyze the available data to determine that they are adequate to support the environmental descriptions of Section 5.3, *Description of the Affected Environment*, and the impact analyses of Section 5.4, *Environmental Impacts*. The following factors should be considered in the analysis:

- Location and number of monitoring stations (and wells) as required to consider the following factors:

    - Meteorological, soil, surface water, and ground water system characteristics in the site vicinity [e.g., surface-water flow fields in the site vicinity, ground water flow (e.g., saltwater intrusion)].

- Impact of sanitary and chemical waste-retention methods on ground water quality:

    - Type of sanitary and chemical waste-retention system; and
    - Transient hydrological and meteorological parameters in the site vicinity.

- Sampling frequency and times to ensure that important temporal variations (e.g., tidal variations and intense rainfall) are adequately monitored.

For review of on-site meteorological instrumentation, the analysis should ensure that the basic meteorological parameters measured by instrumentation include wind direction and wind speed at two elevations, and ambient air temperature difference between two elevations. Guidance on meteorological data to be used as input to atmospheric dispersion modeling and assessment is given in Regulatory Guides 1.111 (NRC, 1977b) and 1.21 (NRC, 1974). Guidance on instrument types, sampling heights, and locations is given in Regulatory Guide 1.23, Sections C.1 and C.2 (NRC, 1972). Guidance on effluent and environmental monitoring at uranium mills is given in Regulatory Guide 4.14 (NRC, 1980).

Operational Monitoring

The operational monitoring program is designed to establish the impacts of operation of the facility and to detect any unexpected impacts arising from facility operation. Operational monitoring may be required by other permitting agencies.

The environmental PM should verify that sufficient information has been provided to adequately assess the environmental monitoring program (e.g., measuring sediment transport and floodplains or wetlands) to: (i) describe the appropriate local and regional chemical characteristics; (ii) ensure environmental protection; and (iii) provide an adequate database for evaluation of the effects of facility operation.

The following information should be included in the EIS:

- Description of the results of the baseline monitoring program, including monitoring station locations and the methods, frequency, and duration of monitoring used in each case [Tables and maps should be used, if appropriate.];

- Intensity of sampling needed for each anticipated impact. It should be commensurate with the degree of impact expected;

- Validity of data; and

- Adequacy of data measurement techniques.

5.6.3 Ecological Monitoring

This section describes the major components of the applicant's proposed ecological monitoring program. Monitoring programs should cover elements of the ecosystem for which a causal relationship between construction, operation, or decommissioning and adverse change is established or strongly suspected.

The environmental PM should describe the applicant's/licensee's ecological monitoring program. Two aspects of monitoring should be considered:

- Baseline monitoring to support the applicant's baseline descriptions and provide information for operational comparison; and

- Operational monitoring to establish the impacts of operation of the facility and detect any unexpected impacts arising from facility operation.

Each of these aspects is discussed in greater detail below.

Baseline Monitoring

The program of ecological field monitoring is used to support the applicant's descriptions of the ecological environment. Baseline monitoring is needed to establish a database from which to observe potential future impacts. Generally, data are needed on a seasonal basis and should be sufficient to characterize seasonal variations throughout at least one annual cycle. Additional data may be needed on a site-specific basis.

The environmental PM should analyze the available data to determine that they are adequate to support the environmental descriptions of Section 5.3, *Description of the Affected Environment*; and the impact analyses of Section 5.4, *Environmental Impacts*. The following factors should be considered in the analysis:

- The location and number of monitoring stations as required to consider the following factors:

  - Distribution and abundance of "important" species, habitats, and communities [Critical life history information should include parameters such as feeding areas, wintering areas, and migration routes to the extent that the proposed action is expected to affect these parameters.]; and

-   Descriptions of any modifications that may affect the existing patterns of plant and animal communities (e.g., changing agricultural practices, development of holding ponds or reservoirs, and developing access routes).

Operational Monitoring

A program of operational ecological monitoring may be necessary to monitor the environmental impacts of facility or site operation. It continues the studies conducted during pre-operational monitoring. An operational monitoring program may be included with an application for an operating license, and for license renewal applications. Operational monitoring programs may not be fully developed at the time of applying for a construction permit.

When evaluating the ecological monitoring programs, the following features should be considered:

•   Ensure that the applicant/licensee has, to the extent feasible, described the general scope and objectives of its intended programs and has provided a tentative list of parameters that should be monitored. The application should include:

    -   Duration over which the parameters will be monitored; and
    -   Provisions for updating the program.

•   Establish whether adequate data will be provided as outlined above [If the monitoring programs are judged to be inadequate or to include unnecessary elements, the environmental PM should evaluate potential additions and deletions.];

•   Consider the following features for the monitoring programs:

    -   Relationship to environmental monitoring conducted by other agencies in the vicinity of the facility or site should be described;

    -   Basis and objective of each element of the monitoring program should be clearly stated, as well as its relationship to the overall environmental monitoring program;

    -   If outputs of a preceding monitoring program or project demonstrate no significant impacts, then provisions to study such effects in successive monitoring programs should be reduced or deleted;

    -   The program should allow for periodic modification based on the results of previous monitoring to ensure that the current monitoring effort is sufficient and justified when compared to a current assessment of the effects that the proposed action/alternative are having on the environment; and

- Intensity of sampling required for each anticipated impact should be commensurate with the degree of impact expected [The reviewer should balance the potential impacts of any sampling program against the potential benefits when making this evaluation.];

- Measurement and sampling methods (e.g., sampling locations and equipment, the pattern, frequency, and duration of sampling and sample size) should be described;

- Statistical validity, including the mean, standard deviation, confidence limits, and sample size should be clearly indicated; and

- If population dynamics models were used in the impact analyses, determine if sampling data are available to support the model. If not, suggest such sampling if verification of the model is necessary.

The following information should be included in the EIS:

- Description of the results of the baseline monitoring program, including monitoring station locations and the methods, frequency, and duration of monitoring used in each case. Tables and maps should be used if appropriate;

- Intensity of sampling needed for each anticipated impact [Sampling intensity should be commensurate with the degree of impact expected.];

- Validity of data; and

- Adequacy of data measurement techniques.

5.7 Cost-Benefit Analysis

This section describes the major costs and benefits for each alternative. Consideration of the costs and benefits should be presented in the EIS (10 CFR 51.71). The costs and benefits should not be limited to a simple financial accounting of project costs for each alternative. Costs and benefits should also be discussed for qualitative subjects (i.e., environmental degradation or enhancement). Extensive or detailed analysis should be presented in an appendix to the EIS to avoid diverting attention away from primary issues such as public health and safety. The cost-benefit analysis is not simply a mathematical formula from which to justify economic parameters; other applicable qualitative factors should be discussed and weighed in the decision.

The environmental PM should describe the costs and benefits for the proposed action and each alternative. Qualitative environmental costs and benefits can be compared to the discussion of environmental impacts within the environmental report. Standard project costs can be reviewed utilizing standard cost estimating databases. Socioeconomic costs and benefits can be reviewed and compared against similar projects as applicable. NUREG/BR-0058 (NRC, 1995a) provides guidance for determining public health and safety impact valuation. NUREG-1530 (NRC, 1995b) provides background material and information relating to NUREG/BR-0058. The reviewer should also verify that analyses were performed in accordance with appropriate cost benefit guidance. Future costs and benefits should be discounted to present worth as discussed in "Economic Analysis of Federal Regulations Under

Executive Order 12866" found on the WWW at
<http://www.whitehouse.gov/OMB/inforeg/riaguide.html> (OMB, 1996). This site also provides general guidance on calculating costs and benefits. The methods used for discounting should be explained, and applied consistently to both costs and benefits. NUREG–1727, *NMSS Decommissioning Standard Review Plan* (NRC, 2000), provides guidance on determining costs and benefits for decommissioning projects as well as providing guidance on determining ALARA and prohibitive costs related to ALARA.

The cost benefit analysis provides input to determine the relative merits of various alternatives; however, the NRC must ultimately base its decision on public health and safety issues.

5.8   Summary of Environmental Consequences

This section should summarize any adverse environmental impacts that cannot be avoided and for which no practical means of mitigation are available, the relationship between short-term uses of the environment and the maintenance and enhancement of long-term productivity, and any irreversible or irretrievable commitments of resources which would be involved. As appropriate, this summary can be tabulated.

The environmental PM should ensure the following analysis is completed:

- Develop a list of:

    - Unavoidable adverse environmental impacts;

    - Irreversible and irretrievable commitments of resources (those materials that would be irretrievably committed during construction, operation, and decommissioning);

    - Short- or long-term impacts (consider that occupation of land for an indefinite period represents the maximum impact on long-term productivity, unless other long-term preemptions have been identified; identify through consultation with the appropriate reviewers other uses of the environment that will be precluded by facility construction, operation, and decommissioning and classify these as either short-term or long-term preemptions; determine how any short-term or long-term benefits of the proposed action affect any such preemptions.);

    - Procedures and practices to mitigate or avoid these impacts or commitments; and

    - Impacts or commitments that remain after all practical means to avoid or mitigate the impact have been taken;

- Categorize the identified impacts (direct, indirect, and cumulative) according to the resource (e.g., water resource);

- The categories may be further divided into construction, operational, and decommissioning impacts, if so desired;

- Ensure that each identified impact has been appropriately categorized. When a particular action or operation results in multiple impacts (e.g., access road construction and use may have impacts affecting land use, terrestrial ecology, and socioeconomic), ensure that the impacts are addressed in each appropriate category;

- Determine the magnitude of the impacts (direct, indirect, and cumulative) or commitments; and

- Evaluate the time scale of each impact (e.g., 4–6 months during construction, throughout the facility lifetime, indefinitely).

The information from Sections 5.3, *Description of the Affected Environment*; and 5.4, *Environmental Impacts* should be summarized for this section. The EIS includes a discussion of the predicted short-term unavoidable adverse environmental impacts of each alternative and the predicted long-term environmental impacts. Short-term represents the period from start of construction to end of the proposed action, including prompt decommissioning. Long-term represents the period extending beyond the end of the proposed action. The discussion should also include an evaluation of the extent to which the proposed action will preclude options for other future use of the environment. "Irreversible" impacts refer to commitments of environmental resources that cannot be restored. "Irretrievable" applies to material resources and will involve commitments of materials that, when used, cannot be recycled or restored for other uses by practical means. The following information should be listed in the EIS for the proposed action and each alternative:

- Unavoidable adverse environmental impacts;

- Irreversible and irretrievable commitment of resources;

- Short-term and long-term impacts; and

- Short-term uses of the environment and the maintenance and enhancement of long-term productivity.

For new facilities the maximum long-term impact to productivity would result if the facility is not dismantled at the end of the period of facility operation, and consequently the land occupied by the facility structures would not be available for any other use. For operating or decommissioning facilities the maximum long-term impact to productivity would occur if the restricted release criteria are used for decommissioning.

After reviewing the impacts and mitigation actions, organize these impacts by environmental categories and prepare a brief paragraph summarizing the nature and magnitude of each category of impact in sufficient detail to allow for a comparative analyses of the environmental consequences of each alternative. Table 3 illustrates an example format of a table used to describe the nature and magnitude of each impact.

Table 3. Example of environmental impacts

Impact Category	Adverse Impacts Based on Applicant's Proposal	Actions to Mitigate Impacts	Unavoidable Adverse/ Irreversible and Irretrievable Commitments of Resources/Short- and Long-Term Impacts
Regional Setting			
Geology and Soil			
Water Resource			
Ecological			
Air Quality			
Noise			
Historic and Cultural			
Visual/Scenic			
Socioeconomic			
Environmental Justice			
Public and Occupational Health			
Waste Management			

5.9 List of Preparers

This section should contain a list of preparers and credentials who participated in producing the EIS.

5.10 Distribution List

This section should contain a list of all parties to whom the EIS was distributed.

5.11 References Cited

All references used in the preparation of the EIS should be listed, including those cited in the text of the EIS and those that were not specifically cited but served as useful guidance during document development. Additionally, it is helpful to provide ADAMS Accession numbers, if applicable, to assist the public in finding relevant documents. Guidance in NUREG-0650 (NRC, 1999) should be useful for determining reference format.

5.12 Supplemental Information of Environmental Impact Statement Document

Appendices should be included at the end of the EIS that include information that is supportive of the findings in the EIS. Examples include:

- Scoping report;
- Glossary;
- Consultation letters;
- Dose assessments;
- Issues Eliminated from detailed study; and
- Technical evaluations.

5.13 References

American Society for Testing and Materials (ASTM), 1996. "Standard Guide for Selection of Environmental Noise Measurements and Criteria." ASTM E1686. ASTM, West Conshohocken, PA.

Bass, R.E.; Herson, A.I.; and Bogdan, K.M.; 2001. "The NEPA Book: A Step-by-Step Guide on how to Comply with the National Environmental Policy Act." Solano Press Books, Point Arena, CA.

CEQ (Council on Environmental Quality) 1981, "Forty Most Asked Questions Concerning CEQ's National Environmental Policy Act Regulations." CEQ, Executive Office of the President, Washington, D.C. <http://ceq.eh.doe.gov/nepa/regs/40/40P1.HTM >. (December 18, 2002).

EPA (U.S. Environmental Protection Agency), 1974. "Information of Levels of Environmental Noise Requisite to Protect Public Health and Welfare with an Adequate Margin of Safety." EPA 550/9-74-004. EPA, Washington, DC. March.

BLM (Bureau of Land Management) 1984. "Visual Resource Management." BLM Manual 8400. BLM, U.S. Department of Interior, Washington, DC. April. <http://www.blm.gov:80/nstc/VRM/8400.html>. (January 13, 2003).

BLM, 1986a. "Visual Resource Inventory." BLM Manual Handbook H-8410-1. BLM, U.S. Department of Interior, Washington, DC. January. <http://www.blm.gov:80/nstc/VRM/8410.html>. (January 13, 2003).

BLM, 1986b. "Visual Resource Contrast Rating." BLM Manual Handbook H-8431-1. BLM, U.S. Department of Interior, Washington, DC. January 1986b. <http://www.blm.gov:80/nstc/VRM/8431.html>. (January 13, 2003).

BLM, 2003. "U.S. Department of the Interior–Bureau of Land Management Visual Resource Management." <http://www.blm.gov:80/nstc/VRM/index.html>. (January 13, 2003).

NPS (National Park Service), 2002. "How to Apply the National Register Criteria for Evaluation." *Bulletin No. 15*. NPS, U.S. Department of the Interior, Washington, DC. <http://www.cr.nps.gov/nr/publications/bulletins/nrb15/>. (January 13, 2003).

NPS, 2003a. "National Register of Historic Places Publications Page." NPS, U.S. Department of Interior, Washington, DC. <http://www.cr.nps.gov/nr/publications/>. (January 13, 2003).

NPS, 2003b. "NPS Archeology and Ethnography Home Page." NPS, U.S. Department of Interior, Washington, DC. <http://www.cr.nps.gov/aad/>. (March 17, 2003).

NRC (U.S. Nuclear Regulatory Commission), 1972. "Onsite Meteorological Programs (Safety Guide 23) (Draft SS 926-4, Proposed Revision 1, published September 1980) (Draft ES 926-4, Second Proposed Revision 1, published April 1986)". Regulatory Guide 1.23, Revision 1. U.S. Nuclear Regulatory Commission, Washington, DC. April.

NRC, 1974. "Measuring, Evaluating, and Reporting Radioactivity in Solid Wastes and Releases of Radioactive Materials in Liquid and Gaseous Effluents from Light Water-Cooled Nuclear Power Plants." Regulatory guide 1.21, Revision 1. U.S. Nuclear Regulatory Commission, Washington, DC.

NRC, 1975. "Programs for Monitoring Radioactivity in the Environs of Nuclear Power Plants." Regulatory Guide 4.1, Revision 1. U.S. Nuclear Regulatory Commission, Washington, DC. April.

NRC, 1977a. "Final Environmental Statement on the Transportation of Radioactive Material by Air and Other Modes." NUREG–0170. U.S. Nuclear Regulatory Commission, Washington, DC.

NRC, 1977b. "Methods for Estimating Atmospheric Transport and Dispersion of Gaseous Effluents in Routine Release from Light-Water Cooled Reactors." Regulatory Guide 1.111, Revision 1. U.S. Nuclear Regulatory Commission, Washington, DC

NRC, 1980. "Radiological Effluent and Environmental Monitoring at Uranium Mills." Regulatory Guide 4.14, Revision 1. U.S. Nuclear Regulatory Commission, Washington, DC. April.

NRC, 1995a. "Regulatory Analysis Guidelines of the U.S. Nuclear Regulatory Commission." NUREG-0058. U.S. Nuclear Regulatory Commission, Washington, DC. November.

NRC, 1995b. "Reassessment of NRC's Dollar Per Person-Rem Conversion Factor Policy." NUREG–1530. U.S. Nuclear Regulatory Commission, Washington, DC. December.

NRC, 1997. "Questions Related to the Parks Township Environmental Impact Statement." Memorandum to Papereillo from Olmstead. U.S. Nuclear Regulatory Commission, Washington, D.C. October 30.

NRC, 1999. "Preparing NUREG-Series Publications." NUREG–0650. U.S. Nuclear Regulatory Commission, Washington, DC.

NRC, 2000. "NMSS Decommissioning Standard Review Plan." NUREG–1727. U.S. Nuclear Regulatory Commission, Washington, DC. September 2000.

NRC, 2001. "Regulations Handbook." NUREG/BR–0053, Revision 5. U.S. Nuclear Regulatory Commission, Washington, D.C. March.

OMB (Office of Management and Budget), 1996. "Economic Analysis of Federal Regulations Under Executive Order 12866." OMB, Executive Office of the President, Washington, DC. January 11. <http://www.whitehouse.gov/OMB/inforeg/riaguide.html>. (January 13, 2003).

6 THE ENVIRONMENTAL REPORT: FORMAT AND TECHNICAL CONTENT

This chapter provides information on the content of the ER and is also applicable to supplemental ERs.

This chapter generally follows the outline of an EIS as presented in Chapter 5, though an ER may also be used in support of an EA. The applicant/licensee may benefit from a pre-licensing meeting between the licensing PM and the environmental PM to discuss the information needed to support the environmental review (e.g., information normally contained in the ER). The goal of these meetings is to define the scope and detail required within the ER. Chapter 5 describes how the NRC staff uses the ER information to prepare an EIS.

The scope of the ER should be balanced against the credible threat to the environment posed by the proposed action (e.g., facility construction, facility operation, or decommissioning). **The ER should present a detailed and thorough description of each affected resource for evaluation of potential impacts to the environment. It may not be necessary for every resource to receive the same level of detailed review and every action may not require all the information discussed in this chapter. Likewise, the proposed action may present unique issues and require additional information not identified in this chapter.** This is consistent with one of the goals of NEPA, which is to concentrate on issues significant to the proposed action and their potential environmental impacts.

General ER requirements are provided in the NRC implementing regulations for NEPA (e.g., 10 CFR 51.45 for general requirements, 10 CFR 51.54 for manufacturing licenses, 10 CFR 51.60 for materials licenses and 10 CFR 51.62 for 10 CFR 61 disposal sites).

6.1   Introduction of the Environmental Report

The introduction should be brief, and should include a description of the proposed action, a brief summary of pertinent statutes and regulations, location of the proposed action and relevant background information. Key dates and deadlines should also be listed to establish the time frame for the proposed action.

6.1.1 Purpose and Need for the Proposed Action

The applicant/licensee should explain why the proposed action is needed. This section of the ER describes the underlying need for the proposed action and should not be written merely as a justification of the proposed action, nor to alter the choice of alternatives. Another common mistake is to identify compliance with NEPA and CEQ regulations as the need. Examples of need include a benefit provided if the proposed action is granted or descriptions of the detriment that will be experienced without approval of the proposed action. In short, the need describes what will be accomplished as a result of the proposed action.

6.1.2 The Proposed Action

The following information should be presented in the ER, as applicable. It may not be necessary for the evaluation of potential impacts from the proposed action to require all the information requested below:

- Brief description of the proposed action, including the name of the applicant/licensee;

- Regional and site area maps, including nearby towns and natural features;

- Schedule of the major steps comprising the proposed action, such as construction, operation, decommissioning (i.e., start and completion dates of major activities); and

This section should also describe the desired outcome or goal of the proposal. For example, at a decommissioning site, the licensee must meet the 10 CFR 20, Subpart E, radiological criteria for license termination. For a new fuel cycle facility, the applicant/licensee must meet the 10 CFR 70 criteria.

6.1.3 Applicable Regulatory Requirements, Permits, and Required Consultations

For some of these consultations, NRC may designate the applicant/licensee as responsible for performing the consultation process. The following information should be presented in the ER, as applicable. It may not be necessary for the evaluation of potential impacts from the proposed action to require all the information requested below:

- Name of each consultation, review, approval, and authorization, and the applicable law, ordinance, or regulation;

- Activity to be covered by the consultation, review, approval, or authorization (e.g., permit);

- Current status of each consultation, review, approval, and authorization;

- Potential administrative delays or other problems preventing agency consultation, review, approval, or authorization; and

- Documentation of any consultation or survey conducted, such as wildlife surveys (periodic or one-time) or archaeological surveys.

6.2 Alternatives

6.2.1 Detailed Description of the Alternatives

The applicant/licensee should discuss alternatives considered. Identify the no-action alternative, the proposed action, and any reasonable alternatives. Discuss the technical design requirements for the proposed action and the reasonable alternatives. It is possible to have options under an alternative (e.g., the possibility of additional ground water remediation) and those options should be discussed.

6.2.1.1 No-Action Alternative

The applicant/licensee should identify the no-action alternative in order to provide a baseline to compare the proposed action and reasonable alternatives. The no-action alternative is the status-quo. The following information should be presented in the ER, as applicable. It may not be necessary for the evaluation of potential impacts from the proposed action to require all the information requested below:

- Description of the no-action alternative; and

- Summary of the major impacts should the no-action alternative be chosen.

6.2.1.2 Proposed Action

The applicant/licensee should describe the proposed action as described in the following, as applicable. It may not be necessary for the evaluation of potential impacts from the proposed action to require all the information requested below:

- Detailed description of the proposed action, the general project progression and pre-operational, operational, and post-operations activities, as appropriate;

- Full names of all organizations sharing ownership of the proposed action;

- The major impacts from performing the proposed action;

- Measures used to mitigate impacts;

- Restoration actions; and

- Proposed monitoring.

Additionally, the applicant/licensee should describe the current state of the site or facility. The following information should be presented in the ER, as applicable:

- Site location, including distance and direction from the nearest major city, nearby towns, nearby inhabitants, and landmarks, including highways, rivers, or other bodies of water;

- Facility latitude and longitude coordinates;

- Areal extent of the site and facility layout;

- The following maps which include the facility area and scale of the map:

    - Sufficiently detailed map showing highways and railroad lines that cross the site;

    - Aerial view or perspective drawing of the site with an indication of the facility boundary (in at least one drawing the facility site boundary should occupy about 10 percent of the view);

- Layout of facilities and other features within the site boundary with the same scale as those provided for Section 6.4, *Environmental Impacts*;

- List of buildings or areas used for chemical storage, waste management, vehicle cleaning, administration, operations and maintenance, generating electricity, health and security, parking, etc.;

- Underground storage tanks, wells, pipelines, and sewage system;

- Description of types of operations that will be conducted on the site;

- Identification of radionuclides and other hazardous materials used;

- Summary of how materials are stored, handled, utilized and disposed; and

- Air, ground water, and surface water, monitoring stations.

6.2.1.3 Reasonable Alternatives

The applicant/licensee should summarize the history and process used to formulate the reasonable alternatives. The following information should be provided for each reasonable alternative, as applicable. It may not be necessary for the evaluation of potential impacts from the proposed action to require all the information requested below:

- A description of the alternative;

- The major impacts;

- Measures used to mitigate impacts;

- Restoration and management actions; and

- Proposed monitoring.

6.2.2 Alternatives Considered but Eliminated

The following information should be presented in the ER, as applicable. It may not be necessary for the evaluation of potential impacts from the proposed action to require all the information requested below:

- Summary of alternatives not considered to be reasonable; and

- Summary of why these alternatives were eliminated from further study.

6.2.3 Cumulative Effects

Discuss any past, present, or reasonably foreseeable future actions which could result in cumulative impacts when combined with the proposed action.

6.2.4 Comparison of the Predicted Environmental Impacts

The applicant/licensee should present the impacts of the proposed action and alternatives in a summary chart or table.

6.3 Description of the Affected Environment

The description of the affected environment focuses on baseline conditions, i.e., the status quo. The baseline conditions will be used to assess the impacts discussed in Section 6.4, *Environmental Impacts*.

The following information should be presented in the ER, as applicable. It may not be necessary for the evaluation of potential impacts from the proposed action to require all the information requested below:

- Land use;
- Transportation;
- Geology and soils;
- Water resources;
- Ecology;
- Meteorology, climatology, and air quality;
- Noise;
- Historical and cultural resources;
- Visual/scenic resources;
- Socioeconomic;
- Environmental justice;
- Public and occupational health; and
- Waste management.

6.3.1 Land Use

The applicant/licensee should describe land uses near the site. This section provides input to various sections including, but not limited to, Sections 6.4.1, *Land Use Impacts*; 6.4.4, *Water Resources Impacts*; 6.4.12, *Public and Occupational Health Impacts*; and 6.6, *Environmental Measurements and Monitoring Program*.

The following information should be presented in the ER, as applicable. It may not be necessary for the evaluation of potential impacts from the proposed action to require all the information requested below:

- Maps showing major land use, public, and trust land areas;

- Description of the regional setting, transportation corridors, and offsite areas;

- Land areas devoted to major uses according to U.S. Geological Survey land use categories;

- Information from the U.S. Department of Agriculture Natural Resources Conservation Service on the relative value of the facility if it involves farmland;

- Land-use plans including current, future, and proposed (those which have been formally proposed by the appropriate governing body in a written form and are being actively pursued by officials of the jurisdiction) plans;

- Staged plans, which must go through phases of development, including those that are incomplete;

- Special land-use classifications (e.g., American Indian or military reservations, wild and scenic rivers, State and national parks, national forests, designated coastal zone areas, wildlife refuges, wilderness areas, and prime and unique farmlands);

- Mineral resources;

- Principal agricultural products, location, and average annual yields (including growing and grazing period, fraction of daily intake from pasture, fraction of the year that leafy vegetables are grown, and amount consumed);

- Present commercial fish and invertebrate catch; and

- Unusual animals, facilities, agricultural practices, game harvests, or food processing operations.

6.3.2 Transportation

The applicant/licensee should describe transportation facilities at and near the site. This section provides input to various sections including, but not limited to, Sections 6.4.7, *Noise Impacts* and 6.4.12, *Public and Occupational Health Impacts*.

The following information on existing transportation corridors should be presented in the ER, as applicable. It may not be necessary for the evaluation of potential impacts from the proposed action to require all the information requested below:

- Proposed routes for transportation corridors that will be used for transportation access to and from the facility site; and

- Corridor lengths, widths, and areas including:

  - Identification of offsite transportation areas by land use, size, and location; and
  - Land use restricting transportation corridors contained in any easements.

6.3.3 Geology and Soils

The applicant/licensee should identify the geological, seismological, and geotechnical characteristics of the site and vicinity. This section provides input to various sections including, but not limited to, Sections 6.4.3, *Geology and Soils Impacts*; 6.4.4, *Water Resources Impacts*; 6.4.12., *Public and Occupational Health Impacts*; and 6.6, *Environmental Measurements and Monitoring Program*.

The following information should be presented in the ER, as applicable. It may not be necessary for the evaluation of potential impacts from the proposed action to require all the information requested below:

- Stratigraphy and structures, including descriptions of geological units, major structural and tectonic features (e.g., faults), and any other significant geological conditions;

- Geotechnical investigations conducted to characterize the site;

- Characteristics of soil, including a physical description of the soil units and descriptions of features related to soils at the site and nearby; and

- Analysis and evaluation of the local and regional seismicity data, volcanism, or any information that may indicate a geologic hazard at the site.

6.3.4 Water Resources

The applicant/licensee should describe site-specific and regional data on the physical and hydrological characteristics of ground and surface water in sufficient detail to provide the basic data for the evaluation of impacts on water bodies, aquifers, aquatic ecosystems, and social and economic structures of the area. This section provides input to various sections including, but not limited to, Sections 6.4.4, *Water Resource Impacts*; 6.4.12, *Public and Occupational Health Impacts*; and 6.6, *Environmental Measurements and Monitoring Program*.

The following information should be presented in the ER, as applicable. It may not be necessary for the evaluation of potential impacts from the proposed action to require all the information requested below:

- Maps showing:

  - The spatial and temporal relationship of the site to the major surface and subsurface hydrological systems such as aquifer systems and drainage basins;

  - Surface and subsurface systems that could be affected by facility withdrawals and/or discharges (cross sections where feasible);

- Mean, range, and temporal and spatial variations of the subsurface and surface water quality characteristics including water temperature, chemical, biological, and physical characteristics typically monitored [WWW at <http://www.epa.gov/storet>, (EPA, 2003a)];

- Descriptions of preexisting environmental conditions and their effects on subsurface and surface water quality (e.g., water bodies at or near the site that do not meet established water quality standards) and quantity;

- Historical and current hydrological data from non-related projects in the region or area of influence (e.g., reservoirs built and operated during the period of record; scheduled construction of dams; local drinking water, agricultural, or industrial wells), and projected data describing future trends, if available;

- Interpolated and extrapolated measurements using acceptable geostatistical techniques if data are incomplete or unavailable;

- Summary of methodology used to estimate hydrological parameters;

- Water rights and resources;

- Quantitative description of subsurface and surface water uses such as withdrawals, consumption, and returns, including but not limited to, domestic, municipal, agricultural, industrial, mining, recreation, navigation, and hydroelectric power;

- Quantitative and qualitative description of recreational, navigational, instream, and other non-consumptive water uses including the use rate with time variation;

- Descriptions of past, current, and future pollutant sources with discharges to water including locations relative to the site and the affected water bodies, and the magnitude and nature of the pollutant discharges, including spatial and temporal variations;

- Description of wetlands [WWW at <http://www.usace.army.mil/inet/functions/cw/cecwo/reg/techbio.htm> (USACE, 2003)]; and

- Summary of statutory and other legal restrictions relating to water use or specific water-body restrictions on water use imposed by Federal or State regulations.

Surface Water Characteristics for the following categories:

- Freshwater streams, lakes and impoundments, and estuaries and oceans;

- Flood frequency distributions, including levee failures;

- Flood control measures (reservoirs, levees, flood forecasting);

- Location, size, and elevation of outfall;

- Velocity distribution (horizontal and vertical) and waterbody cross section within the influence of any outfall;

- Bathymetry near any outfall;

- Estimated erosion characteristics and sediment transport for surface-water bodies and wetlands, including rate, bed, suspended load fractions, and gradation analyses;

- Description of the floodplain and its relationship to the site [WWW at <http://www.fema.gov/mit/tsd/> (FEMA, 2003)]; and

- Description of the design-basis flood elevation; and, where applicable, the design-basis flood discharge.

Freshwater streams (for the watershed containing the site):

- Major streams, size of drainage areas, and gradient;

- Historic monthly flow information, including maximum, average-maximum, average, average-minimum, and minimum flow;

- Historical drought stages and discharges by month, and the 7-day once-in-10-yr low flow; and

- Important short-duration flow fluctuations (e.g., diurnal release variations from peaking operation of upstream hydroelectric project).

Lakes and impoundments:

- Elevation-area-capacity curves;

- Reservoir operating rules;

- Annual yield and dependability;

- Variations in inflows, outflows, water-surface elevations, and storage volumes and retention times;

- Net loss, including evaporation and seepage;

- Current patterns, including frequency distributions of current speed, direction, and persistence; and

- Temperature distribution (horizontal and vertical) and stratification and seasonal variations of density-induced currents.

Estuaries and oceans:

- Shoreline and bottom descriptions, including seasonal variations due to sediment transport;

- Tidal current patterns (velocities and phases), range, and excursion;

- Non-tidal circulation patterns, including frequency distributions of current speed, direction, and persistence;

- Temperature and salinity distribution (horizontal and vertical), including temporal variations; and

- Monthly river discharge including maximum, average-maximum, average, average-minimum, and minimum discharge and flushing characteristics (only for estuaries).

Ground water characteristics:

- Historical and seasonal trends in ground water elevation or piezometric levels;

- Piezometric contour maps, water table contour maps, and hydraulic gradients (historical, if available, and current);

- Depth to water table for unconfined aquifer systems;

- Flow travel time (ground water velocities);

- Soil properties, including permeabilities or transmissivities, storage coefficients or specific yields, total and effective porosities, clay content, and bulk densities;

- Interactions among different aquifers;

- Historical and current data from site wells (e.g., monitoring, background, corrective action, or other uses);

- Hydrostratigraphy of the site, including cross sections and hydrostratigraphic unit descriptions; and

- Qualitative description of ground water aquifers, including identification of EPA-designated sole-source aquifers [WWW at <http://www.epa.gov/OGWDW/swp/sumssa.html> (EPA, 2003b)].

6.3.5 Ecological Resources

The applicant/licensee should describe species types, spatial and temporal distribution, and abundance, especially as they relate to listed and endangered species and critical habitat. This section provides input to various sections including, but not limited to, Sections 6.4.5, *Ecological Resources Impacts*; 6.4.12, *Public and Occupational Health Impacts*; and 6.6, *Environmental Measurements and Monitoring Program.*

The following information should be presented in the ER, as applicable. It may not be necessary for the evaluation of potential impacts from the proposed action to require all the information requested below:

- Map(s):

  - Important terrestrial resources, habitats, and ecosystems; and
  - Topographic maps of the site;

- General ecological description of the regional setting, the site, and transportation corridors;

- List and description of important species and their spatial and temporal distributions, including their relative abundance and their life histories, critical life stages, biologically significant activities, seasonal habitat requirements and population fluctuations, food chain, and other interspecific relationships;

- List of threatened or endangered species (plants and animals) known to occur, or that could potentially occur, including their seasons of occurrence, estimates of abundance, local flight patterns, and critical habitats;

- List of major vegetation layers (e.g., over-story and under-story), their dominant species, and the relative species abundances;

- Qualitative estimate of the importance of habitat of threatened, endangered, and other important species relative to the habitat of such species throughout their entire range;

- Locations of travel corridors for important terrestrial species and alternate routes for those corridors that could potentially be blocked by use of the site;

- List of important ecological systems that are especially vulnerable to change or that contain important species habitats, such as breeding areas (e.g., nesting areas), nursery, feeding, resting, and wintering areas, or other areas of seasonally high concentrations of individuals of important species;

- Characterization of the aquatic environment (including biological, hydrological, and chemical) and identification of those factors known to influence distribution and abundance of threatened and endangered aquatic life;

- Location and value of the commercial and sport fisheries and the seasonal distribution of harvest by species;

- Key aquatic indicator organisms expected to gauge changes in the distribution and abundance of species populations that are particularly vulnerable to impacts from the proposed action;

- List of important ecological systems onsite or in the vicinity that are especially vulnerable to change or that contain important species habitats, such as breeding areas (e.g., spawning areas); nursery, feeding, and wintering areas; or other areas of seasonally high concentrations of individuals of important species;

- Relative significance of various aquatic habitats in a regional context;

- Description of current and reasonable foreseeable conditions that are indicative of ecological stresses including natural and man-made;

- Description of the status of ecological succession of biota (i.e., weed, brush, pole, and mature stages);

- Description and location of any ecological or biological studies of the site or its environs, including those that are currently in progress;

- Information on sightings of endangered or threatened species on the proposed site or in the applicable vicinity (e.g., county, tri-county area, bay area, etc.) [The source of this information should be identified. Example sources may include the State Department of Natural Resources, local chapters of recognized bird-watching groups, documented field studies, and State university/college specialists. The time period in which the information was collected by these sources should be specified (e.g., during the past 5 years of monthly observation outings.];

- Documentation that the applicant has consulted with the appropriate Federal and State agencies (e.g., as required by the Fish and Wildlife Coordination Act) and affected American Indian tribes; and

- Identification of other Federal and State projects within the region that are or could potentially affect the same threatened and endangered species or their habitats.

6.3.6 Meteorology, Climatology, and Air Quality

The applicant/licensee should characterize atmospheric transport and diffusion processes at and near the site of the proposed action. This section provides input to various sections including, but not limited to, Sections 6.4.6, *Air Quality Impacts*; 6.4.12, *Public and Occupational Health Impacts*; and 6.6, *Environmental Measurements and Monitoring Program*.

The following information should be presented in the ER, as applicable. It may not be necessary for the evaluation of potential impacts from the proposed action to require all the information requested below:

Meteorology and Climatology

- Description of the general climate of the region (e.g., climatological normals of parameters such as temperature, precipitation, and wind speed/direction);

- Discussion of the severe weather phenomena (e.g., tornadoes, hurricanes, thunderstorms, atmospheric stagnation episodes) experienced in the region with expected frequencies of occurrence and measured extremes of parameters such as temperature, precipitation, and wind speed;

- Monthly and annual air temperature and dewpoint temperature summaries (including averages, measured extremes, and diurnal range) as near as possible to the site;

- Monthly and annual summaries of precipitation, including averages and measured extremes, number of hours with precipitation, and hourly rainfall rate distribution as near as possible to the site;

- Description of the local airflow patterns and characteristics, using data collected from the onsite meteorological program or from nearby weather monitoring stations;

- Description of the baseline air quality in the region, identifying pollutants which are in non-attainment or maintenance areas and the relationship of the site to these areas;

- Monthly and annual wind roses and wind direction persistence summaries at all heights at which data on wind characteristics are applicable centered on the site, if possible;

- Hourly averages of wind speed and direction at all heights which wind characteristics are applicable and a summary of atmospheric stability;

- Estimated monthly mixing height data, including frequency and duration of inversion conditions and methods used to provide the estimates; and

- Topographic data presentation, including a map showing detailed topographic features.

If appropriate meteorological data are not available for the site, applicable data from nearby sources may be used if sufficient justification for offsite data is provided. Information sources for the above information include:

- Onsite meteorological program data;

- National Weather Service stations, [WWW at <http://www.nws.noaa.gov/> (NOAA, 2003a)];

- National Environmental Data Index, (WWW at <http://www.nedi.gov/> (NOAA, 2003b); or

- National Climatic Data Center, (WWW at <http://www.ncdc.noaa.gov/> (NOAA, 2003c).

Baseline Air Quality

- General description of regional air quality, sources of information include:

    - EPA Air Quality System [WWW at <http://www.epa.gov/ttn/airs/airsaqs/> (EPA, 2003c)]; and

    - EPA Aerometric Information Retrieval System [WWW at <http://www.epa.gov/air/data/index.html> (EPA, 2003d);

- Table comparing regional air quality parameters to National Ambient Air Quality Standards for the area, if possible;

- Air pollutants for which there are non-attainment or maintenance areas in the region and a map relating the site to these areas; and

- Local or regional emission inventory.

6.3.7 Noise

The applicant/licensee should characterize the noise baseline at and near the site of the proposed action. This section may require input from various sections including, but not limited to, 6.2.1.2, *Proposed Action*; 6.3.1, *Land Use*; 6.3.6, *Meteorology, Climatology, and Air Quality*; 6.3.10 *Socioeconomic*; and provides input to various sections including, but not limited to, Section 6.4.7, *Noise Impacts*.

The following information should be presented in the ER, as applicable. It may not be necessary for the evaluation of potential impacts from the proposed action to require all the information requested below:

- Boundaries of the extent of the noise analysis;

- Distribution of people, buildings, roads, and recreational facilities that are vulnerable to noise impacts by the proposed action;

- Current and historical noise levels at sensitive areas, as identified above, as energy equivalent sound level (L_{eq}) or the day-night average sound level (L_{dn}) reported on the dBA scale;

- Topography and land use in the vicinity;

- Meteorological conditions in the vicinity;

- Applicable sound level standards (from consultation with Federal, State, regional, local, and affected American Indian tribal agencies); and

- Point and line sources of noise affecting current noise levels.

6.3.8 Historic and Cultural Resources

The applicant/licensee should identify and describes historic, archaeological, and cultural resources. Resources can include districts, sites, buildings, structures, or objects. This section provides input to various sections including, but not limited to, Section 6.4.8, *Historic and Cultural Resources Impacts*.

The following information should be presented in the ER, as applicable. It may not be necessary for the evaluation of potential impacts from the proposed action to require all the information requested below:

- Extent of historical and cultural resource analyses;

- Known cultural resources in the area and an overview of the area's cultural setting;

- Archaeological or historical surveys of the proposed site, including the following:

  - Physical extent of the survey (if the entire site was not surveyed, the basis for selecting the area to be surveyed is needed);

- Brief description of the survey techniques used and the reason for the selection of the survey techniques used;

- Qualifications of the surveyors; and

- Findings of the survey in sufficient detail to permit a subsequent independent assessment of the impact of the proposed project on archaeological and historic resources;

- List of cultural and historic properties within the proposed actions site or within the area of potential effects [These properties are included in State or local registers or inventories of historic and archaeological resources. Guidance can be found on the US. National Park Service WWW at <http://www.cr.nps.gov/nr/publications> (NPS, 2003)];

- The results of any consultation with Federal, State, local, and affected American Indian tribal agencies;

- The comments from any organizations and individuals contacted by the applicant who provided significant information concerning the location and assessment of cultural and historic properties; and

- Statement of the significance or importance of each cultural resource potentially affected.

6.3.9 Visual/Scenic Resources

The applicant/licensee should provide information on the aesthetic and scenic quality of the site, which provides input to various sections including, but not limited to, Sections 6.4.9, *Visual/Scenic Resources Impacts* and 6.4.10, *Socioeconomic Impacts*.

The following information should be presented in the ER, as applicable. It may not be necessary for the evaluation of potential impacts from the proposed action to require all the information requested below:

- Boundaries of the viewshed or viewscape of the proposed action;

- Photos viewing the proposed site from different directions;

- Identification of local residents and/or regular visitors to the area who might be affected by aesthetic impacts;

- Information related to the landscape characteristics including open spaces, mountain ranges, ecological environment (e.g., flora, fauna, and ecosystems), bodies of water (e.g., waterfalls, waterways, and oceans), color of soils, recreational areas (e.g., parks wilderness areas), architectural features, aesthetic (e.g., historical, archaeological, cultural, and natural) features that would attract tourists, and uncultivated land;

- Location of constructed features including radar towers, transmission towers, and overhead power distribution line and production activities;

- Visibility from access roads (i.e., existing natural or constructed barriers, screens or buffers);

- Regionally or locally important or high quality views associated with proposed action sites;

- Photos and information related to the view of the proposed action from different directions including views from roads, highways, homes, and recreational areas (e.g., forest and wilderness area and campgrounds);

- Regulatory information related to land-use zoning requirements of the local community or urban areas, sign ordinances or regulations of the local community or urban area, design guides of the local community or urban area, and buffer-zone (or greenbelt-zone) requirements of the local community or urban area;

- Summary of any coordination with appropriate local area community planners and/or urban planners; and

- Rating of the aesthetic and scenic quality of the site in accordance with the BLM Visual Resource Inventory and Evaluation System (BLM, 1984, 1986a, 1986b, 2002).

6.3.10 Socioeconomic

The applicant/licensee should describe socioeconomic information. This section provides input to various sections including, but not limited to, Sections 6.4.10, *Socioeconomic Impacts*; 6.4.11, *Environmental Justice*; and 6.4.12, *Public and Occupational Health Impacts*. This section may also be linked to Sections 6.3.1, *Land Use*, 6.3.9, *Visual/Scenic Resources*, 6.4.1, *Land Use Impacts*, and 6.4.9, *Visual/Scenic Impacts* because of land use questions.

The following information should be presented in the ER, as applicable. It may not be necessary for the evaluation of potential impacts from the proposed action to require all the information requested below:

- Population characteristics (e.g., ethnic groups, and population density);

- Economic trends and characteristics, including employment and income levels;

- Housing, health and social services, educational, and transportation resources;

- Area's tax structure and distribution;

- Summary of any coordination with appropriate local and regional agencies or groups who collect these types of data;

- Map identifying places of significant population grouping, such as cities and towns;

- Population characteristics (trends) and projections [sources of information include the WWW at <http://www.census.gov> (CB, 2003)] and the bases for population projections;

- Areas where minority or low-income populations are disproportionately high (see Environmental Justice instructions in Appendix C); and

- Sources of information, assumptions and techniques used to develop information.

Current and projected population levels for the life of the facility should be determined. The population trends at the proposed site should be discussed along with historic and projected growth rates for the region. Appropriate governmental and industrial projections should be evaluated. Any unusual programs or developments in the region should be highlighted if they may have an impact on the area population.

6.3.11 Public and Occupational Health

The applicant/licensee should describe existing public and occupational health issues, as appropriate. This section provides input to various sections including, but not limited to, Section 6.4.12, *Public and Occupational Health Impacts*.

The following information should be presented in the ER, as applicable. It may not be necessary for the evaluation of potential impacts from the proposed action to require all the information requested below:

• Major sources and levels of background radiation exposure, including natural and man-made sources; express levels in mSv/yr (mrem/yr);

• Current sources and levels of exposure to radioactive materials;

• Major sources and levels of chemical exposure; express levels in appropriate units;

• Historical exposures to radioactive materials;

• Occupational injury rates and occupational fatality rates; and

• Summary of health effects studies.

6.3.12 Waste Management

The applicant/licensee should describe current waste generation rates and sources for all types of waste. This section provides input to various sections including, but not limited to, Section 6.4.13, *Waste Management Impacts*. This section may be linked to Sections 6.4.1, *Land Use Impacts*; 6.4.4, *Water Resources Impacts*; 6.4.5, *Ecological Resources Impacts* 6.4.6; *Air Quality Impacts*; 6.4.12.2.1, *Pathway Assessment*; and 6.6, *Environmental Measurements and Monitoring Programs*.

The following information should be presented in the ER, as applicable. It may not be necessary for the evaluation of potential impacts from the proposed action to require all the information requested below:

• Descriptions of all (i.e., nonradioactive, radioactive, mixed, and hazardous) current waste systems, including quantities, composition, and frequency of waste generation [Effluent discharges do not need to be discussed if previously covered (i.e., air effluents in Air Quality section and liquid effluents in the Water Quality section)];

• Information on current disposal activities including size and location of disposal sites as well as the plans for ultimate treatment and/or restoration of retired disposal sites (other than licensed commercial sites);

- Identification of all sources of radioactive liquid, solid, and gaseous waste material within the facility; and

- Identification of direct radiation sources stored onsite as solid waste (e.g., independent fuel storage).

6.4   Environmental Impacts

Analyze and describe the impacts for each resource described in Section 6.3, *Description of the Affected Environment*, for the no-action alternative, the proposed action, and each alternative. These impacts (e.g., direct, indirect, and cumulative) should consider normal operational events as well as reasonably foreseeable accidents (e.g., design basis events for 10 CFR 72 licensees, credible consequence events for 10 CFR 70 licensees). When analyzing impacts, resources should be considered separately, and where necessary, in combination (e.g., noise impacts on wildlife, or transportation impacts on land use), as appropriate

Activities (i.e., construction, operation and decommissioning) should be evaluated in sufficient detail to determine the significance of potential impacts and to recommend how these impacts should be treated in the process (e.g., consideration of alternative designs or practices that would mitigate adverse environmental impacts).

6.4.1   Land Use Impacts

This section describes the impacts to land use for each alternative. The following information should be presented in the ER, as applicable. It may not be necessary for the evaluation of potential impacts from the proposed action to require all the information requested below:

- Land-use impact;

- Land-use impacts of any related Federal action that may have cumulatively significant impacts;

- Area and location of land that will be disturbed on either a long-term or short-term basis; and

- Impacts from institutional controls.

6.4.2   Transportation Impacts

This section describes the impacts to transportation corridors including the effects of transportation of radioactive materials. The following information should be presented in the ER, as applicable. It may not be necessary for the evaluation of potential impacts from the proposed action to require all the information requested below:

- Construction of access road or railroad to facility;

- Transportation route and mode for conveying construction material to the facility;

- Traffic pattern impacts (e.g from any increase in traffic from heavy haul vehicles);

- Impacts of construction transportation such as fugitive dust, scenic quality, and noise;

- Mitigation measures proposed by applicant; and

- Any consultations with Federal, State, and local agencies.

<u>Transportation of Radioactive Material</u>

The following information should be provided in the ER:

- Transportation mode (i.e., truck, rail, or barge) and routes from originating site to destinations;

- Estimated transportation distance from the originating site to the storage site;

- Treatment and packaging procedure for radioactive wastes;

- Radiological dose for incident-free scenarios to public and workers; and

- Impacts of operating transportation on the environment (e.g., fire from equipment sparking).

6.4.3 Geology and Soils Impacts

This applicant should summarize known and potential geological impacts, mitigation measures and cumulative effects in this section. The major analysis for this section is usually found in the SER and only needs to be summarized in this section. Examples of geological environmental impacts include soil compaction, soil erosion, subsidence, landslides, and disruption of natural drainage patterns. More likely, geological resources may exert an impact on the proposed action and these effects should be summarized in this section (e.g., seismic or volcanic hazards).

6.4.4 Water Resources Impacts

In this section, the applicant/licensee evaluates impacts on water use and water quality for each alternative and identifies the potential impacts for both radiological and nonradiological effluents.

The applicant should consider surface-water and ground water uses that could affect or be affected by the construction and operation of the proposed project. The analysis includes consideration of impacts on such water uses as domestic, municipal, agricultural, industrial, mining, recreation, navigation, and hydroelectric power. The review should be limited to present and known future water uses.

Consider impacts on the physical, chemical, and biological water-quality characteristics of ground and surface water. Because water quality and water supply are interdependent, changes in water quality must be considered simultaneously with changes in water supply.

Compliance with environmental quality standards and requirements of the Clean Water Act is not a substitute for and does not negate the requirement for the applicant to weigh the environmental impacts of the proposed action, including any degradation of water quality, and to consider alternatives to the proposed action that are available for reducing the adverse impacts. Additionally, the State's standards should be considered because the United States Supreme Court granted the States additional authority to limit hydrological alterations beyond the State's role in regulating water rights.

The following information should be presented in the ER, as applicable. It may not be necessary for the evaluation of potential impacts from the proposed action to require all the information requested below:

- Identification of waters receiving effluents and the expected average and maximum flow rates, physical characteristics (e.g., temperature, sediment load, velocities), and composition of radiological and nonradiological pollutants in these effluents;

- Impacts on surface water and ground water quality including comparison of predicted effluent and receiving-water quality with applicable effluent limitations and water-quality standards for both radiological and nonradiological constituents [Include conclusions regarding project compliance with these standards, the physical impacts of consumptive water uses (e.g., ground water depletion) on other water users, and adverse impacts on surface-oriented water users (e.g., fishing, navigation, etc.) resulting from facility activities];

- Identification of hydrological system alterations, including construction of cofferdams and storm sewers; dredging operations; placement of fill material into the water; creation of shoreline facilities involving bulkheads, piers, jetties, basins, or other structures or activities with potential to alter existing shoreline processes; construction of intake and outfall structures; water-channel modifications; construction of roads and bridges; operations affecting water levels (flooding); dewatering activities; and activities contributing to sediment runoff (e.g., road construction, clearing and grading, fill or spoil placement);

- Identification of hydrological system impacts, onsite and offsite (e.g., water quantity and availability, water flow, and movement patterns), and erosion, deposition, and sediment transport, water drainage characteristics, the flood handling capability of the floodplains, flow and circulation patterns, subsidence resulting from ground water withdrawal, and erosion and sediment transport;

- Withdrawals and returns of ground and surface water during all phases;

- Identification of impacted ground and surface water users, including descriptions of the site and regional water bodies (including sole-source aquifers) and ground water aquifers (Section 5.3.5, *Water Resources*), surface-water and ground water sources used, identification and locations of ground water and surface water users and areas that could be impacted, the compatibility of proposed water uses with existing and known water rights and allocations, descriptions of any transfer of water rights (e.g., from irrigation use to facility consumptive use) and the impacts associated with such transfers;

- Descriptions of any proposed practices and measures to control impacts to water quality and/or quantity (e.g., protection of natural drainage channels and water bodies, protection of shorelines and beaches, restrictions on access to and use of surface water, protection against saltwater intrusion, and handling of fuels, lubricants, oily wastes, chemical wastes, sanitary wastes, herbicides, and pesticides); and

- Identification of predicted cumulative effects on water resources.

6.4.5 Ecological Resources Impacts

This section describes the ecological impacts for the proposed action and each alternative. The following information should be presented in the ER, as applicable. It may not be necessary for the evaluation of potential impacts from the proposed action to require all the information requested below:

- Site map showing proposed buildings, land to be cleared, areas to be cleared along stream banks, areas proposed for dredge material, areas to be dredged, and waste disposal areas;

- Documentation of Section 7 consultations with the FWS on the impact of the proposed action on endangered and threatened species and critical habitat, as discussed in Section 1.4;

- Proposed schedule of activities;

- Total area of land to be disturbed;

- Area of disturbance for each habitat type, and an estimate of the amount of these habitats that will be destroyed relative to the total amount present in the region;

- Maintenance practices such as use of chemical herbicides, roadway maintenance, and mechanical clearing that are anticipated to effect biota;

- Area to be used on a short-term basis during construction, and plans for restoration of this land;

- Any proposed activities expected to impact communities or habitats that have been defined as rare or unique or that support threatened and endangered species;

- Estimate of the potential impacts of elevated construction equipment or structures on species (e.g., birds collisions, nesting);

- Tolerances and/or susceptibilities of important biota to physical and chemical pollutants;

- Clearing methods, erosion, run-off and siltation control methods (both temporary and permanent), dust suppression methods, and other construction practices for impact control or minimization;

- Special maintenance practices used in important habitats (e.g., marshes, natural areas, bogs) including those that result in unique beneficial effects on specific biota;

- Wildlife management practices; and

- Practices and procedures or alternative designs to minimize adverse impacts.

6.4.6 Air Quality Impacts

This section describes the air quality impacts of the proposed action and each alternative. The following information should be presented in the ER, as applicable. It may not be necessary for the evaluation of potential impacts from the proposed action to require all the information requested below:

- Description of gaseous effluents (type, quantity, and origin);

- Table comparing effluent concentrations to regional air quality parameters (effluent concentrations should be provided for both short and long term impacts);

- Release point characteristics (i.e., elevation above grade, inside vent or stack diameter, physical shape, flow rate, effluent temperature, exit velocity, release frequency, or other appropriate information to allow calculation of transport and diffusion);

- Description of gaseous effluent control systems;

- Detailed descriptions of the models and assumptions used to determine normalized concentration and/or relative deposition [The meteorological data used in these models should be identified (Section 6.3.6, *Meteorology, Climatology, and Air Quality*).];

- Normalized concentration and/or relative deposition at points of potential maximum concentration outside the site boundary, at points of maximum individual exposure, and at points within a reasonable area that could be impacted (Section 6.3.6, *Meteorology, Climatology, and Air Quality*);

- Description of visibility impacts;

- Description of mitigative measures for air quality impacts; and

- Description of cumulative air quality impacts.

6.4.7 Noise Impacts

This section describes noise impacts. The following information should be presented in the ER, as applicable. It may not be necessary for the evaluation of potential impacts from the proposed action to require all the information requested below:

- Predicted noise levels (sound contour maps are recommended), reported as energy equivalent sound levels or day-night average sound levels (L_{eq} or L_{dn}) using the dBA scale;

- Major point and line sources (for locations described above), including all models, assumptions and input data;

- Comparison to appropriate standards or guidelines (EPA, 1974; ASTM, 1996);

- Potential impacts to sensitive receptors (i.e., hospitals, schools, residences, wildlife);

- Mitigation measures to reduce impacts of noise; and

- Description of noise related cumulative impacts.

6.4.8 Historic and Cultural Resources Impacts

This section describes impacts to historic and cultural resources. Adverse effects occur when a proposed action's effect on a cultural resource diminishes the integrity of its location, design, setting, materials, workmanship, feeling or association. Adverse effects include, but are not limited to: (i) physical destruction, damage, or alteration of all or part of the property; (ii) isolation of the property from or alternation of the character of the property's setting when that character contributes to the property's qualification of the *National Register*; (iii) introduction of visual, audible or atmospheric elements that are out of character with the property or alter its setting; (iv) neglect of a property resulting in its deterioration or destruction; and (v) transfer, lease or sale of the property.

The following information should be presented in the ER, as applicable. It may not be necessary for the evaluation of potential impacts from the proposed action to require all the information requested below:

- Overlay maps where a base map showing known and potential sites is overlain by maps identifying the nature and extent of the impacts from each alternative [Summary information that does not include site-specific or property-specific data should be included in cases where specific information may lead to vandalism or scavenging.];

- Impacts to historic and cultural resources during construction, operation, or decommissioning;

- Indirect impacts (e.g., vandalism on known cultural resource sites in the area of potential effects, visual impact, denial of access) resulting from land-use changes, secondary growth and development, or direct construction activities;

- Documentation of SHPO and/or THPO consultations on the impact of the proposed action on significant cultural and historic resources as discussed in Section 1.4;

- Reference to SHPO and/or THPO comments on the impact of the proposed project on significant cultural and historic resources as discussed in Section 1.4;

- State laws and plans for historic preservation, if needed;

- Potential for human remains to occur in the project area and plans for complying with Native American Graves Protection and Repatriation Act regulations in the event of an inadvertent discovery [An inadvertent discovery of such items during construction may necessitate a work stoppage of up to 30 days and consultation under this Act's procedures.];

- Practices and procedures or alternative designs to minimize adverse impacts [Mitigation measures could include: (i) limiting the magnitude of the undertaking; (ii) modifying the undertaking through redesign, reorientation or construction on the proposed action; (iii) repair, rehabilitation, or restoration of an affected historic property as opposed, for instance, to demolition; (iv) preservation and maintenance operations for involved historic properties; (v) documentation (drawings, photos, histories) of building or structures that must be destroyed or substantially altered; (vi) relocation of historic properties; and (vii) salvage of archaeological or architectural information and materials.]; and

- Description of cumulative impacts on historic and cultural resources.

6.4.9 Visual/Scenic Resources Impacts

This section describes aesthetic impacts. The following information should be presented in the ER, as applicable. It may not be necessary for the evaluation of potential impacts from the proposed action to require all the information requested below:

- Photos of the site with the alternatives superimposed;

- Rate the aesthetic and scenic quality of the site in accordance with BLM Visual Resource Management System (BLM, 1984, 1986a, 1986b, 2002);

- Significant visual impacts from each alternative, including;

    - Physical facilities that are out of character with overall existing architectural features;

    - Structures that may partially or completely obstruct views of existing landscape;

    - Structures that create visual intrusions in the existing landscape character (e.g., radar towers, power lines, etc.);

    - Structures that may require the removal of natural or built barriers, screens or buffers, thus enabling lower quality viewscapes to be seen;

    - Altering historical, archaeological or cultural properties or the character of the property's setting when that character contributes to the property's significance; and

    - Structures that create visual audible or atmospheric elements that are out of character with the site or alter its setting;

- A determination if the visual impact is compatible or in compliance with regulations, ordinances, and requirements;

- Potential mitigation measures; and

- Description of cumulative impacts to visual/scenic quality.

6.4.10 Socioeconomic Impacts

This section describes socioeconomic impacts such as impacts to housing or schools from an influx of additional workforce. The following information should be presented in the ER, as applicable. It may not be necessary for the evaluation of potential impacts from the proposed action to require all the information requested below:

- Impacts to population characteristics (e.g., ethnic groups, and population density);

- Impacts to housing, health and social services, educational, and transportation resources;

- Impacts to area's tax structure and distribution;

- Summary of any coordination with appropriate local and regional agencies or groups who collect these types of data;

- Sources of information, assumptions and techniques used to develop information; and

- Description of cumulative impacts to socioeconomic resources.

6.4.11 Environmental Justice

The Commission has directed the staff to develop an environmental justice (EJ) policy statement. After the policy statement is completed, necessary updates to the EJ guidance will be incorporated. In the interim, the following draft guidance on environmental justice is being provided.

This section evaluates environmental impacts on low-income or minority populations by proposed project activities if disproportionately high low-income or minority populations are identified in Section 6.3.11. Impacts that may have environmental justice implications may include health, ecological (including water quality and water availability), social, cultural, economic and aesthetic resources.

The ER should follow the detailed guidance provided in Appendix C. In general, the ER includes a discussion of the methods used to identify and quantify impacts on low-income and minority populations, the location and significance of any environmental impacts during construction on populations that are particularly sensitive, and any additional information pertaining to mitigation.

The following information should be presented in the ER, as applicable. It may not be necessary for the evaluation of potential impacts from the proposed action to require all the information requested below:

- An assessment (qualitative or quantitative, as appropriate) of the degree to which each minority or low-income population is disproportionately receiving adverse human health or environmental impacts during construction or decommissioning as compared with the entire geographic area [In addition, there should be an assessment comparing the impacts with the larger overall geographic area encompassing all of the alternative sites.];

- An assessment (qualitative or quantitative, as appropriate) of the significance or potential significance of such environmental impacts on each low-income and minority population [Significance is determined by considering the disproportionate exposure, multiple-hazard, and cumulative hazard conditions.];

- An assessment of the degree to which each low-income and minority population is disproportionately receiving any benefits compared with the entire geographic area;

- A discussion of any mitigative measures for which credit is being taken to reduce environmental justice concerns;

- A brief description of pathways by which any environmental impacts may result in disproportionate environmental impacts to low-income and minority populations;

- Description of cumulative impacts to low-income and minority populations; and

- When alternative sites are being evaluated, the same reviews should be available for each site.

6.4.12 Public and Occupational Health Impacts

This section describes public and occupational health impacts from both nonradiological and radiological sources.

6.4.12.1 Nonradiological Impacts

The following information should be presented in the ER, as applicable. It may not be necessary for the evaluation of potential impacts from the proposed action to require all the information requested below:

- Maps, in an appropriate scale, showing the distances from the proposed action to the following points or areas for radial sectors centered on the cardinal compass directions:

  - Nearest site boundary;

  - Nearest full time resident;

  - Nearest present drinking water intake (from Sections 6.3.1, *Land Use*, or 6.3.4, *Water Resources*); and

  - Nearest sensitive receptors (e.g., schools and hospitals);

- For liquid nonradioactive discharge to water or air, provide the basis for analysis and the following information (Section 6.4.4, *Water Impacts* and Section 6.4.6, *Air Quality Impacts*):

  - Transit time to the points of analysis;
  - Liquid stream discharge rate; and
  - Dilution factor at the points of analysis;

- Physical layout, including the location and orientation of nonradioactive materials that are expected to be present (Section 6.1.2, *The Proposed Action* and 6.3.12, *Waste Management*);

- Location and characteristics of liquid and gaseous releases (from Sections 6.4.4, *Water Resources Impacts*, and 6.4.6 *Air Quality Impacts*);

- Measured nonradiological concentrations, airborne and waterborne, at specific locations where environmental monitoring data exist (Section 6.6, *Environmental Measurements and Monitoring Programs*);

- Calculated airborne and waterborne concentrations at specific locations important to exposure calculations where environmental monitoring data are not available, including a description of the methodology;

- Calculated exposure to the public or calculated average annual concentration of nonradioactive releases to air and water; including all models, assumptions, and input data in order to determine compliance (e.g, 40 CFR 50, 59, 60, 61, 122, 129, 131, etc.);

- Number and principal locations of workers who will be exposed to the sources described above and the total amount of time per year that they will spend at those locations;

- Calculated exposure to the workforce including all models, assumptions, and input data in order to determine compliance with 29 CFR 1910;

- Description of mitigative measures; and

- Description of nonradiological cumulative impacts to public and occupational health.

6.4.12.2 Radiological Impacts

This section describes public and occupational health impacts from radiological sources.

6.4.12.2.1 Pathway Assessment

The following information should be presented in the ER, as applicable. It may not be necessary for the evaluation of potential impacts from the proposed action to require all the information requested below:

- Maps, at an appropriate scale, showing the distances from the proposed action to the following points or areas for radial sectors centered on the cardinal compass directions:

  - Nearest site boundary;

  - Nearest full time resident;

  - Location of average member of critical group;

  - Other important receptors (i.e: milk and meat producing animals, and vegetable gardens) and locations;

  - Nearest present and known future locations from which an individual can obtain aquatic food and/or drinking water (Sections 6.3.5, *Water Resources* and 6.3.2, *Land Use*), transit time from the proposed action, and population served; and

  - Nearest present and known future areas designated for recreational purposes (Section 6.3.2, *Land Use*) and transit time from the proposed action;

- Potential pathways for releases;

- For each radioactive discharge to water or air, provide the basis for analysis and the following information (Sections 6.4.4, *Water Resources Impacts* and 6.4.5, *Air Quality Impacts*):

  - Transit time to the points of analysis;
  - Discharge rate; and
  - Dilution factor at the points of analysis;

- Distributional data for radial sectors centered on the cardinal compass directions for radial distances (immediate area to affected region) including:

6-27

- Projected population during and after each alternative (Section 6.4.10, *Socioeconomic Impacts*);

- Current annual meat production, current annual milk production, current annual vegetable production, and current commercial fish and invertebrate catch (Section 6.3.2, *Land Use*); and

- Affected current and known future drinking water intake locations and the populations served and the daily water consumption at each location (Section 6.3.5, *Water Resources*);

- Crop yield, annual production, growing period, crop type, and amounts consumed and fractional ingestion of contaminated food and water for:

  - Irrigated land using water withdrawn within the affected region of the proposed action, include irrigation rate; and

  - Land affected by airborne emissions and deposition;

- Animal husbandry, facilities, agricultural practices, game harvests, or food processing operations having the potential for contributing incrementally to either individual or population doses.

6.4.12.2.2 Public and Occupational Exposure

The following information should be presented in the ER, as applicable. It may not be necessary for the evaluation of potential impacts from the proposed action to require all the information requested below:

- Physical layout of the site, including the location and orientation of radioactive materials that are expected to be present (Section 6.2.1.2, *Proposed Action*);

- Location and characteristics of radiation sources and liquid and gaseous radioactive effluent (Sections 6.4.4, *Water Resource Impacts*, and 6.4.6 *Air Quality Impacts*);

- Measured radiation dose rates, airborne radioactivity concentrations, and waterborne radioactivity concentrations at specific locations where environmental radiological monitoring data exist;

- Calculated radiation dose rates, airborne radioactivity concentrations, and waterborne radioactivity concentrations at specific locations important to dose calculations where environmental radiological monitoring data are not available, including a description of the methodology;

- Calculated total effective dose equivalent to an average member of the critical group or calculated average annual concentration of radioactive material in gaseous and liquid effluent; including all models, assumptions, and input data in order to determine compliance with 10 CFR 20 and 40 CFR 190;

- Calculated dose to the workforce including all models, assumptions, and input data in order to determine compliance with 10 CFR 20;

- Summary of external radiation monitoring and airborne radiation monitoring programs (Section 6.6, *Environmental Measurements and Monitoring Programs*);

- Description of mitigation measures; and

- Description of cumulative impacts to public and occupational radiological exposure.

For accidents, include:

- The list of reasonably foreseeable (i.e., credible) accidents (e.g., design basis events for 10 CFR 72 licenses, credible consequence events for 10 CFR 70 licenses, etc.) identified as having a potential for releases to the environment and the analysis of the dose consequences from these accidents.

6.4.13 Waste Management Impacts

This section describes waste generation and management impacts. The following information should be presented in the ER, as applicable. It may not be necessary for the evaluation of potential impacts from the proposed action to require all the information requested below:

- Descriptions of the sources, types, quantities, composition of solid, hazardous, radioactive and mixed wastes expected from the proposed action;

- Description of proposed waste management systems designed to collect, store, and dispose of all wastes generated by the proposed action;

- Anticipated disposal plans for all wastes (i.e., transfer to an offsite waste disposal facility, treatment facility, or storage onsite);

- A waste-minimization plan that identifies process changes that can be made to reduce or eliminate waste, including a description of methods to minimize the volume of waste; and

- Description of waste management cumulative impacts.

6.5   Mitigation Measures

The ER should summarize mitigation measures that could reduce adverse impacts. These mitigation measures should be incorporated in the proposed action and alternatives (40 CFR 1502.14(f) and 1508.20). The anticipated effectiveness of these mitigation measures should be addressed in reducing adverse impacts. Residual impacts or unavoidable adverse impacts which remain after mitigation measures have been applied should be analyzed, as well as any further impacts caused by the mitigation measures themselves. The technical feasibility and the cost-benefit of any potential mitigation measures including costly actions that would yield only minor environmental benefits, should be noted.

6.6   Environmental Measurements and Monitoring Programs

This section describes all environmental measurement and monitoring programs as they apply to baseline, operation, and decommissioning conditions for the proposed action and each alternative.

6.6.1 Radiological Monitoring

The following information should be presented in the ER, as applicable. It may not be necessary for the evaluation of potential impacts from the proposed action to require all the information requested below:

- Maps or aerial photographs of the site with proposed monitoring and sampling locations clearly identified along with effluent release points;

- Principal radiological exposure pathways (Section 6.4.12.2.1, *Pathway Assessment*);

- Location and characteristics of radiation sources and radioactive effluent (liquid and gaseous, from Sections 6.4.4, *Water Resource Impacts*, and 6.4.6, *Air Quality Impacts*);

- Detailed description of the monitoring program including:

  - Number and location of sample collection points, measuring devices used, and pathway sampled or measured;

  - Sample size, sample collection frequency, and sampling duration; and

  - Method and frequency of analysis including lower limits of detection;

- Discussion justifying the choice of sample locations, analyses, frequencies, durations, sizes, and lower limits of detection; and

- Quality assurance procedures.

6.6.2 Physiochemical Monitoring

The following information should be presented in the ER, as applicable. It may not be necessary for the evaluation of potential impacts from the proposed action to require all the information requested below:

- Maps or aerial photographs of the site clearly identifying proposed monitoring and sampling locations, effluent release points, and parameter being measured/analyzed;

- Chemical parameters (e.g., nitrogen dioxides or particulates from an industrial off-gas discharge unit, chlorides or pH from a wastewater outfall);

- Physical parameters (e.g., wind speed and direction, temperature, precipitation, etc.);

- Detailed description of the monitoring program including:

  - Number and location of sample collection points, measuring devices used, and pathway sampled or measured;

  - Sample size, sample collection frequency, and sampling duration;

- Method and frequency of analysis including lower limits of detection;

- Discussion justifying the choice of sample locations, analyses, frequencies, durations, sizes, and lower limits of detection;

- Quality assurance procedures;

- Description of action levels and corrective action requirements;

- Physical parameters (e.g., air temperature, wind speed, ground water levels, surface water flow rates); and

- Map showing detailed topographic features of the site (as modified by the facility), including major structures and the meteorological tower/s (if applicable).

6.6.3 Ecological Monitoring

The following information should be presented in the ER, as applicable. It may not be necessary for the evaluation of potential impacts from the proposed action to require all the information requested below:

- Maps showing features of the site and transportation corridors that will be modified, including major ecological communities, important habitats, and sampling stations and monitoring locations;

- List and description of the important ecological resources that are likely to be affected;

- List of monitoring program elements or parameters including action or reporting levels for each element;

- Type, frequency, and duration of observations or samples taken at each location, and appropriate rationale and sampling design;

- Statistical validity of any existing or proposed sampling program [For quantitative descriptions of samples collected within each area of interest and each time of interest, descriptive statistics should include: the mean, standard deviation, standard error, and confidence interval for the mean. In each case, the sample size should be clearly indicated. If diversity indices are used to describe a collection of organisms, the specific diversity indices used should be stated. Also, the methods used for observing natural variations of ecological parameters should be described. If these methods involve indicator organisms, the criteria for their selection should be stated.];

- Sampling equipment used;

- Method of chemical analyses, as applicable;

- Data analysis and reporting procedures;

- Documentation of applicant consultations with the FWS, appropriate State agencies (e.g., fish and wildlife agency), and American Indian tribal agencies; and

- Documentation of the environmental monitoring programs in policy directives designating a person or organizational unit responsible for reviewing the program on an ongoing basis.

Procedures should establish criteria for (as applicable):

- Data recording and storage;

- Reporting results to the NRC or consulting agency; and

- Actions to be taken for anomalous results or when results do not meet requirements.

6.7   Cost Benefit Analysis

This section describes the costs and benefits for the proposed action and each alternative. NUREG/BR-0058 and NUREG-1530 (NRC, 1995a; 1995b) provide detailed guidance. The discussion of costs and benefits will include both the costs of each alternative and a qualitative discussion of environmental impacts. Provide assumptions and uncertainties in the analyses.

The following information (major costs and benefits) should be presented in the ER, as applicable. It may not be necessary for the evaluation of potential impacts from the proposed action to require all the information requested below:

- Qualitative discussion of environmental degradation (including air, water, soil, biotic, as well as socioeconomic factors such as noise, traffic congestion, overuse of public works and facilities, and land access restrictions);

- Decreased public health and safety;

- Capital costs of the proposed action and alternatives, including land and facilities;

- Operating and maintenance costs;

- Post-operation restoration (not applicable when the alternative is restoration);

- Post-operation monitoring requirements;

- Other costs of the alternative (e.g., lost tax revenue, decreased recreational value, degradations in transportation corridors. as appropriate);

- Qualitative discussion of the environmental benefits;

- Increased public health and safety;

- Capital benefits of the alternative;

- Tax revenues received by local, State, and Federal governments;

- Incremental increases in regional productivity;

- Enhancement of recreational values;

- Creation and improvement of transportation corridors and facilities; and

- Other benefits.

6.8 Summary of Environmental Consequences

The following information should be presented in the ER, as applicable. It may not be necessary for the evaluation of potential impacts from the proposed action to require all the information requested below:

- Unavoidable adverse environmental impacts;

- Irreversible and irretrievable commitments of resources used in project construction, operation, and decommissioning;

- Short-term and long-term impacts; and

- Short-term uses of the environment and the maintenance and enhancement of long-term productivity.

6.9 List of References

To be completed by the applicant/licensee indicating items referenced in ER.

6.10 List of Preparers

To be completed by the applicant/licensee indicating personnel completing the ER.

6.11 References

American Society for Testing and Materials (ASTM), 1996. "Standard Guide for Selection of Environmental Noise Measurements and Criteria." ASTM E1686. ASTM, West Conshohocken, PA..

BLM (Bureau of Land Management) 1984. "Visual Resource Management." BLM Manual 8400. U.S. Department of Interior, Washington, DC. April. <http://www.blm.gov:80/nstc/VRM/8400.html>. (January 13, 2003).

BLM, 1986a. "Visual Resource Inventory." BLM Manual Handbook H–8410–1. U.S. Department of Interior, Washington, DC. January. <http://www.blm.gov:80/nstc/VRM/8410.html>. (January 13, 2003).

BLM, 1986b. "Visual Resource Contrast Rating." BLM Manual Handbook H–8431–1. U.S. Department of Interior, Washington, DC. January 1986b. <http://www.blm.gov:80/nstc/VRM/8431.html>. (January 13, 2003).

BLM, 2003. "U.S. Department of the Interior–Bureau of Land Management Visual Resource Management." U.S. Department of Interior, Washington, DC. <http://www.blm.gov:80/nstc/VRM/index.html>. (January 13, 2003).

CB (U.S. Census Bureau), 2003. "Census Bureau Homepage." U.S. Department of Commerce, Washington, DC. <http://www.census.gov>. (March 14, 2003).

FEMA (Federal Emergency Management Agency), 2003. "FEMA: Flood Hazard Mapping." Federal Emergency Management Agency, Washington, DC. <http://www.fema.gov/mit/tsd/>. (February 25, 2003).

EPA (U.S. Environmental Protection Agency), 1974. "Information of Levels of Environmental Noise Requisite to Protect Public Health and Welfare with an Adequate Margin of Safety." EPA 550/9-74-004. EPA, Washington, DC. March.

EPA, 2003a. "EPA > Water > Wetlands, Oceans, & Watersheds > Monitoring and Assessing Water Quality > STORET." EPA, Washington, DC. <http://www.epa.gov/storet/>. (February 25, 2003).

EPA, 2003b. "EPA National Summary of Sole Source Aquifer Designations." EPA, Washington, DC. <http://www.epa.gov/OGWDW/swp/sumssa.html>. (February 25, 2003).

EPA, 2003c. "EPA - TTN AIRS AQS-Air Quality System." EPA, Washington, DC. <http://www.epa.gov/ttn/airs/airsaqs/>. (March 14, 2003).

EPA, 2003d. "EPA AirData - Access to Air Pollution Data." EPA, Washington, DC. <http://www.epa.gov/air/data/index.html>. (March 14, 2003).

NOAA (National Oceanic and Atmospheric Administration), 2003a. "NOAA - National Weather Service." U.S. Department of Commerce, Washington, DC. <http://www.nws.noaa.gov/>. (March 14, 2003).

NOAA, 2003b. "NEDI (National Environmental Database Index) - Homepage." U.S. Department of Commerce, Washington, DC. <http://www.nedi.gov/>. (March 14, 2003).

NOAA, 2003c. "NCDC - National Climatic Data Center (NCDC) Page." U.S. Department of Commerce, Washington, DC. <http://www.ncdc.noaa.gov/>. (March 14, 2003).

NPS (National Park Service), 2003. "National Register of Historic Places Publications Page." NPS, U.S. Department of Interior, Washington, DC. <http://www.cr.nps.gov/nr/publications/>. (January 13, 2003).

NRC (U.S. Nuclear Regulatory Commission), 1995a. "Regulatory Analysis Guidelines of the U.S. Nuclear Regulatory Commission." NUREG-0058. U.S. Nuclear Regulatory Commission, Washington, DC. November.

NRC, 1995b. "Reassessment of NRC's Dollar Per Person-Rem Conversion Factor Policy." NUREG–1530. U.S. Nuclear Regulatory Commission, Washington, DC. December.

OMB (U.S. Office of Management and Budget), 1996. "Economic Analysis of Federal Regulations Under Executive Order 12866" OMB: Washington, DC. <http://www.whitehouse.gov/OMB/inforeg/riaguide.html> (January 13, 2003).

USACE (U.S. Army Corps of Engineers), 2003. "Technical and Biological Information." U.S. Department of Defense, Washington, DC. <http://www.usace.army.mil/inet/functions/cw/cecwo/reg/techbio.htm>. (February 25, 2003).

APPENDIX A
APPLICATION OF THE GENERIC ENVIRONMENTAL IMPACT STATEMENT ON THE LICENSE TERMINATION RULE TO ENVIRONMENTAL ASSESSMENTS FOR UNRESTRICTED RELEASE DECOMMISSIONING SITES

PAGE INTENTIONALLY BLANK

License Termination Rule
GEIS Reference Facilities[1,2] Checklist

The Generic Environmental Impact Statement (GEIS) reference facilities were developed to broadly and generically represent categories of licensee facilities. Specific facilities will not exactly match the descriptions of the reference facilities. The primary purpose of comparing a specific facility to the reference facility with regard to dose assessment is to determine whether the specific facility has important contaminants, potential scenarios, or pathways that were not analyzed for the reference facilities or which may be sufficiently different from those in the GEIS to change conclusions regarding environmental impacts. In general, if a specific facility has contaminants, concentrations, and spacial distributions less than or generally equivalent to those used for the reference facilities, the GEIS should be applicable. Potential limitations of the GEIS dose assessments, as well as a summary of the characteristics of the reference facilities, are shown below.

1.   GEIS Dose Assessment Scenarios: Potential Limitations

     a.   Building Occupancy (structures)

          i.   Structures are assumed to have a 70-year life span following license termination. A shorter expected life span is acceptable. Expected life spans significantly longer than 70 years may require additional analysis if long-lived radionuclides are involved.

          ii.   If Radon (Rn-222) due to licensee activities or co-mingled material is expected to approach or exceed the EPA guideline of 4 pCi/l indoor air concentration, additional dose assessment may be required.

          iii.   Contamination significantly more extensive than that analyzed in the GEIS should be evaluated on a site-specific basis. Areas and concentrations analyzed in the GEIS are shown in the tables in the following sections.

          iv.   Radionuclides present on the site that contribute significantly to dose but which were not analyzed in the GEIS for the subject facility type will need to be evaluated separately.

Checklist for Structures

Yes   No
☐   ☐   Additional analysis required due to expected >70 year building lifespan following decommissioning <u>and</u> long-lived contaminants.
☐   ☐   Indoor Radon (Rn-222) concentration expected to approach or exceed the EPA guideline of 4 pCi/l.
☐   ☐   Contamination significantly more extensive than that shown in Tables 1 through 6 in the following sections.
☐   ☐   Radionuclides present that contribute significantly to dose, were not analyzed in the GEIS, and could change the conclusions in the GEIS regarding environmental impacts.

---

[1]Overview from NUREG-1496, Volume 1, Section 3

[2]Note: The GEIS does not apply to uranium mills or tailings, low level waste, or high level waste.

b.   Residential (soil)

   i.   Assumes people live and work on site over a 1,000 year period.

   ii. If the site is subject to weather or other events (tornadoes, flash floods, etc) that could result in extensive redistribution or mass movement of contaminates, additional analysis may be required.

   iii. Pre-existing contamination of ground water must be evaluated on a site-specific basis.

   iv. 10 CFR 20.302/20.2002 or other burials or disposal areas may need additional site-specific evaluation.

Checklist for Soil

Yes	No	
☐	☐	Site subject to weather or other events that could redistribute contaminants in ways not analyzed in the GEIS.
☐	☐	Contaminated groundwater present.
☐	☐	On-site burials or disposal areas.

2.   Example fuel cycle facilities: power, test, and research reactors; uranium fuel fabrication; uranium hexafluoride conversion facilities; and independent spent fuel storage installations (ISFSI).

The power, test, and research reactors, and the ISFSI have been consolidated into a single analysis in the GEIS based on common radionuclide contaminants (^{60}Co and ^{137}Cs), and are represented by the analysis for the power reactor.

The uranium fabrication facility is used as the reference for both the fabrication and hexafluoride facilities.

Facility Characteristics Applicable to Dose Modeling

1. Soil Surface Activities for the Radionuclides of Interest[1]	
Radionuclide	Surface Concentration (pCi/g)
Co-60	60
Cs-137	20
Uranium	1,000

[1] From NUREG-1496, Table C.7.1.2

Reference Facility	Structures Radionuclide Activity[2], dpm/100 cm²	Structures Surface Areas				Soil Surface Area, ft²	
		ft²		% Contaminated			
		Floor	Wall	Floor	Wall	Total Site	Contaminated
PWR	7.5 x 10⁶ Co60 2.4 x 10⁶ Cs137	250,000	300,000	10	2	50 x 10⁶	3,000
Uranium Fuel Fab	18,000 U	240,000	240,000	50	5	4.7 x 10⁶	100,000

2. Total and Contaminated Surface Areas for Structures and Soils at Reference Sites[1]

(1) The estimated surface areas listed above (reproduced from NUREG-1496, Appendix C, Table C.7.1.1) are based on limited information and in many cases represent an engineering judgment based on the size of the building structural facilities and types of operation. These estimates are considered to be conservatively large, i.e., they probably overestimate the actual areas involved.

(2) Radionuclide activity shown is for building surfaces. Radionuclide activity for soil surfaces is given below.

3. Contamination Distribution Used in the GEIS[1]

Reference Facility	Soil Area	Soil Depth	Soil Volume	Below-Building Soil Depth	Below-Building Soil Volume
	ft²	cm	m³	cm	m³
Nuclear Power Plant	3,000	4 - 100	12 - 250	3 - 21	15 - 100
Uranium Fuel Fabrication	100,000	44 - 300	4,000 - 28,000	18 - 29	82 - 129

[1] From NUREG-1496, Appendix C, Attachment C, Table C.1.10 and C.2.6

3. Example Non-Fuel-Cycle facilities: universities; medical institutions; sealed source manufactures; industrial users of radioisotopes; research and development laboratories; and rare metal refineries.

The sealed source manufactures and R&D laboratories are consolidated into a single analysis. The analysis of the rare metals processing facility is used to represent all other non-fuel-cycle facilities with low to medium to significant contamination.

Materials licensees who use only sealed sources or short-lived radioactive materials are not expected to require decontamination of buildings or soil, and therefore the impacts and costs of decommissioning are expected to be minimal. The GEIS does not include a detailed analysis of these licensees. If a licensee in this category does require more extensive analysis, the applicability of the GEIS should be evaluated by comparison to the other non-fuel-cycle reference facilities based on the radioisotopes and contamination levels involved.

Facility Characteristics Applicable to Dose Modeling

4. Total and Contaminated Surface Areas for Structures and Soils at Reference Sites[1]							
Reference Facility	Structures Radionuclide Activity[2], dpm/100 cm^2	Structures Surface Areas				Soil Surface Area, ft^2	
		ft^2		% Contaminated			
		Floor	Wall	Floor	Wall	Total Site	Contaminated
Sealed Source Manufacturer	102,000 Co60 33,300 Cs137	6,000	4,600	10	5	40,000	5,000
Rare Metal Extraction	18,000 Thorium	150,000	180,000	40	10	740,000	100,000

(1) The estimated surface areas listed above (reproduced from NUREG-1496, Appendix C, Table C.7.1.1) are based on limited information and in many cases represent an engineering judgment based on the size of the building structural facilities and types of operation. These estimates are considered to be conservatively large, i.e., they probably overestimate the actual areas involved.

(2) Radionuclide activity shown is for building surfaces. Radionuclide activity for soil surfaces is shown below.

5. Soil Surface Activities for the Radionuclides of Interest[1]	
Radionuclide	Surface Concentration (pCi/g)
Co-60	60
Cs-137	20
Thorium	200

[1] From NUREG-1496, Table C.7.1.2

6. Contamination Distribution Used in the GEIS[1]					
Reference Facility	Soil Area	Soil Depth	Soil Volume	Below-Building Soil Depth	Below-Building Soil Volume
	ft^2	cm	m^3	cm	m^3
Sealed Source	5,000	4 - 90	20 - 425	3 - 21	0 - 2
Rare Metals Extraction	100,000	10 - 60	1,000 - 5,700	0 - 2	0 - 6
	Slag Pile Volume: 7,000 m^3				

[1] From NUREG-1496, Appendix C, Attachment C, Table C.3.6 and C.4.6

APPENDIX B
CATX CHECKLIST

PAGE INTENTIONALLY BLANK

CATX Checklist

Action Name:

Action Location:

Action Description:

CATX Category:

	YES	NO	Need Data
A. Is the action consistent with the Statements of Consideration for the categorical exclusion chosen?			
B. Is the action likely to significantly affect any aspect of the natural environment?			
C. Is the action likely to significantly affect any aspect of the cultural environment including those that might be related to environmental justice?			
D. Is the action likely to generate a great deal of public interest about any environmental issue?			
E. Is there a high level of uncertainty about the action's environmental effects?			

CONCLUSION:

☐ 1. The action is a CATX and requires no further environmental review.

☐ 2. The action is a CATX but requires additional documentation, see Section 2.1, NUREG-1748, attach documentation.

☐ 3. The action requires an EA.

☐ 4. The action requires an EIS.

_____
Licensing Project Manager   Date

B-3

PAGE INTENTIONALLY BLANK

INSTRUCTIONS

- Action Name: Give the project name and license and docket number.

- Action Location: For actions with specific or general locations, give the address. In other cases it may be necessary to provide other information such as county or township name or latitude/longitude.

- Action Description: Be as brief as possible but provide sufficient information to determine the CATX category in which the proposed action fits and to complete the rest of the checklist.

- CATX category: List the categorical exclusion applicable to the action.

The following checklist consists of questions about the likelihood that a particular kind of environmental consequence will result from the proposed action. The licensing PM may consult with technical staff and EPAB, as necessary.

Based on internal review, external review (where appropriate), and research, check "YES," "NO," or "NEED DATA" for each question. Attach documentation as needed to support the answer. If the "NEED DATA" box is checked, the licensing PM may consult with the EPAB PM about what data are needed and/or how to get it. The following sections provide considerations when completing the checklist.

Checklist Question A: Is the action inconsistent with the Statements of Consideration for the categorical exclusion chosen?

Refer to the Statements of Consideration for the final rule (Chapter 2, NUREG-1748 reproduces the majority of this text). Evaluate the proposed action to determine if the text from the final rule considered the proposed action as a categorical exclusion. If the proposed action is not consistent (i.e., outside the envelope of what was considered), check the YES box; or if it is unclear, an EPAB PM should be consulted.

Checklist Question B: Is the action likely to significantly affect any aspect of the natural environment?

Consider whether the proposed action is likely to affect/alter:

- Endangered or threatened species, or it's critical habitat;
- Natural ecosystems;
- Water supplies of humans, animals, or plants;
- Wetlands; or
- Any other environmental media or resource not listed above.

Checklist Question C: Is the action likely to significantly affect any aspect of the cultural environment including those that might be related to environmental justice?

Consider possible impacts on historic, cultural, and scientific resources. Consider whether the action is likely to have physical, visual, or other effects on:

B-5

- Districts, sites, buildings, structures and objects that are included in the *National Register*, or a State or local register of historic places;
- Places of traditional cultural value for Native American group or other community;
- Known archeological sites, or land identified by archeologists consulted by NRC as having high potential to contain archeological resources;
- Any practice of religion (e.g., by impeding access to a place of worship); or
- Minority or low income groups that are disproportionate with its impacts on other groups. Consider unique practices or lifestyles (e.g., subsistence fishing in a pollutant discharge area) and any adverse economic impacts on such a group.

Checklist Question D: Is the action likely to generate a great deal of public interest about any environmental issue?

Consider whether the proposed action is likely to generate a great deal of public interest. If so, consider whether this public interest is likely to focus on environmental issues. For example, a proposal for research and development activities near a National Park area may be of interest to national activist groups, but this is not an environmental issue unless it can be reasonably argued that the location of this proposed action will generate effluents or have some other impact on the natural or cultural environment. Public interest can be raised on a number of issues: impacts on historic buildings, archeological sites, and other cultural resources.

Checklist Question E: Is there a high level of uncertainty about the proposed action's environmental impacts?

Consider whether there is anything not known about the proposed action's potential impacts, and then whether this information gap has any significance. For example, when considering installation of monitoring equipment, it might not be known whether there are archeological sites in the vicinity. If the installation will result in ground disturbance, this uncertainty should be resolved before proceeding with the installation. If the installation will not result in ground disturbance, there may be no need to resolve the uncertainty.

CONCLUSIONS:

1. Consider whether the action clearly qualifies for the CATX listed. If so, check.

2. Consider whether the action requires additional documentation to verify that the action is consistent with Statements of Consideration. If so, attach the additional documentation (see Section 2.1, NUREG-1748).

3. If the action, even with additional documentation, is not consistent with the Statements of Consideration, or if additional special circumstances are present, as indicated by a YES to questions B-E, an EA should be prepared.

4. Usually, an EA will be prepared first, however, if the provisions of 10 CFR 51.20 apply, check this box and notify EPAB.

APPENDIX C
ENVIRONMENTAL JUSTICE PROCEDURES

PAGE INTENTIONALLY BLANK

NOTE: The environmental justice (EJ) guidance contained in this appendix is provided as draft for interim use. The Commission has directed the staff to develop an EJ policy statement. After the policy statement is completed, necessary updates to the EJ guidance will be incorporated.

ENVIRONMENTAL JUSTICE IN NMSS NEPA DOCUMENTS

I. BACKGROUND

On February 11, 1994, The President signed Executive Order 12898 "Federal Actions to Address Environmental Justice in Minority Populations and Low-Income Populations" which directs all Federal agencies to develop strategies for considering environmental justice in their programs, policies, and activities. Environmental justice is described in the Executive Order as "identifying and addressing, as appropriate, disproportionately high and adverse human health or environmental effects of its programs, policies, and activities on minority populations and low-income populations." On December 10, 1997, the Council on Environmental Quality (the Council or CEQ) issued, "Environmental Justice Guidance Under the National Environmental Policy Act." The Council developed this guidance to, "...further assist Federal agencies with their National Environmental Policy Act (NEPA) procedures." As an independent agency, the Council's guidance is not binding on the NRC; however, the NRC considered the Council's guidance on environmental justice in this procedure.

II. POLICY

This procedure provides guidance to the Office of Nuclear Materials Safety and Safeguards (NMSS) staff on conducting environmental justice reviews for proposed actions as part of NRC's compliance with NEPA. This guidance does not create any new substantive or procedural NEPA related requirements. The guidance is merely intended to improve internal NMSS management by helping to ensure that NRC is fully discharging its existing NEPA responsibilities.

It is the policy of NMSS to address environmental justice in every Environmental Impact Statement (EIS) and, as appropriate, supplements to an EIS, that is issued by NMSS. Under most circumstances, no environmental justice review should be conducted where an Environmental Assessment (EA) is prepared. If it is determined that a particular action will have no significant environmental impact, then there is no need to consider whether the action will have disproportionately high and adverse impacts on certain populations. Similarly, the staff should not request public comments on environmental justice issues when a FONSI is concluded. However, in special cases or circumstances, the reviewer may recommend to management that staff conduct an environmental justice analysis in preparing an EA. Such determinations will be made on a case-by-case basis and only where there is an obvious potential that the consideration of specific demographic information at the site may identify significant impacts that would not otherwise be considered. Management (Branch Chief level) will decide on a case-by-case basis when special cases or circumstances exist that require the staff to perform an environmental justice review for an EA.

The level of discussion on environmental justice will vary based on the circumstances of each action. The actual determination of impacts will not change, but the evaluation and analysis may be expanded. Each EIS or special case EA should contain a section that fully describes the environmental justice review process. Policy implementation guidance is provided in Section III for licensing actions and Section IV for rulemakings.

NOTE: The environmental justice (EJ) guidance contained in this appendix is provided as draft for interim use. The Commission has directed the staff to develop an EJ policy statement. After the policy statement is completed, necessary updates to the EJ guidance will be incorporated.

III. <u>POLICY IMPLEMENTATION FOR LICENSING ACTIONS</u>

A. 1. The first step in evaluating environmental justice potential is to obtain demographic data (census data) for the immediate site area and surrounding communities. Data for the state and county will also be necessary. The demographic data should consist of income levels and minority breakdown. In our experience, the recommended geographic area for evaluating census data is the census block group. The census block group was chosen because the U.S. Census Bureau does not report information on income for blocks, the smaller geographic area, and census tracts are too large to identify minority or low income communities. A minority or low-income community may be considered as either a population of individuals living in geographic proximity to one another or a dispersed/transient population of individuals (e.g., migrant workers) where either type of group experiences common conditions of environmental exposure. For the purpose of this procedure, minority is defined as individual(s) who are members of the following population groups: American Indian and Alaska Native; Asian; Native Hawaiian and Other Pacific Islander; African American (not of Hispanic or Latino origin); some other race; and Hispanic or Latino (of any race). The 2000 Census introduced the multiracial category. Anyone who identifies themselves as white and a minority will be counted as that minority group. In the small number of cases where individuals identify themselves as more than one minority, count that individual in a "Two or More Races" group. Low-income is defined as being below the poverty level as defined by the U.S. Census Bureau (e.g., the U.S. Census Bureau's Current Population Reports, Series P-60 on Income and Poverty).

Guidelines for determining the area for assessment are provided in the following discussion. If the facility is located within the city limits, a radius of approximately 0.6 miles (1 square mile) from the center of the site is probably sufficient for evaluation purposes; however, if the facility itself covers this much area, use a radius that would be equivalent to approximately 0.6 miles from the site. If the facility is located outside the city limits or in a rural area, a radius of approximately 4 miles[3] (50 square miles) should be used. These are guidelines; the geographic scale should be commensurate with the potential impact area, and should include a sample of the surrounding population, e.g., at least several block groups. The goal is to evaluate the "communities," neighborhoods, or areas that may be disproportionately impacted. One source of the census data is the Landview computer software by the U.S. Environmental Protection Agency and the U.S. Department of Commerce, Bureau of the Census. This software is updated after each 10-year census. Other sources include the applicant, local governments, state agencies, or local universities. It is recommended that you utilize the Census Bureau's 10-year census for data on minorities and income level. The reviewer should use the best available information. Present the minority and low-income population data for the block groups, county and state in a table in the EIS or EA.

---

[3]Because of the nature of NMSS facilities a 50 mile radius is not automatically required as is the case for NRR facilities.

NOTE: The environmental justice (EJ) guidance contained in this appendix is provided as draft for interim use. The Commission has directed the staff to develop an EJ policy statement. After the policy statement is completed, necessary updates to the EJ guidance will be incorporated.

2. The next step is to compare the percentage of minority population in the block groups in the area for assessment to the state and county percentages of minority population and to compare the area's percentage of economically stressed households to the state and county percentages of economically stressed households. It is possible that the geographic area could cross county and state lines and this should be considered when making comparisons. If the percentage in the block groups significantly exceed that of the state or county percentage for either minority or low-income population, environmental justice will have to be considered in greater detail. As a general matter (and where appropriate), staff may consider differences greater than 20 percentage points to be significant. Additionally, if either the minority or low-income population percentage exceeds 50 percent, environmental justice will have to be considered in greater detail.

The criteria listed above should only serve as a guideline for determining the presence of a minority or low-income population because demographic data may overlook low-income and minority communities if they constitute a relatively small percentage of the total population in the block group. The staff should continue to supplement the environmental justice analysis with the scoping to identify low-income or minority populations. If it is apparent through interviews, public comment/interest, by investigation, or by other scoping activities, that there is a distinct minority or low-income population that may be adversely affected by the proposed action, then the reviewer should proceed with the environmental justice review even if that population was not identified through the use of demographic data. If no minorities or low-income populations are identified in the potentially affected area or environmental impact area, then document the conclusion. The environmental justice review is complete.

B. Staff should look at the demographics of a site early in the review process. Scoping and public participation are a fundamental part of the NEPA process. Staff's approach will depend on the nature of the regulatory action and the demographics at the proposed location. NMSS staff should ensure that minority and low-income populations are given the opportunity to participate. The NRC's regulations require that any affected Indian tribe be invited to participate in the scoping process for an EIS. During scoping meetings for an EIS, for example, staff will solicit input on environmental issues, and the affected communities should be encouraged to develop and comment on possible alternatives to the proposed agency action. As with any scoping activities under NEPA, the measures staff may consider for increasing participation of minority and low-income populations include outreach through groups such as minority business and trade organizations, schools and colleges, labor organizations, or other appropriate groups. In addition, if a representative(s) of the affected population has been identified such as an officer of an organized local group or community leader, the individual(s) should receive notices of meetings and copies of *Federal Register* notices.

When communicating with the public, NMSS staff should consider disseminating information through alternative media such as translating notices (and other documents) into a language other than English, where appropriate.

The EIS should note whether an environmental justice concern was identified during scoping.

NOTE: The environmental justice (EJ) guidance contained in this appendix is provided as draft for interim use. The Commission has directed the staff to develop an EJ policy statement. After the policy statement is completed, necessary updates to the EJ guidance will be incorporated.

C.   1.   Once it is determined that a site does have a potential for an environmental justice concern, it is then necessary to determine if there is a "disproportionately high and adverse" impact (human health or environmental effect) to the minority or low-income population near the site. Impacts of the proposed action are to be determined in the usual manner, including cumulative and multiple impacts, where appropriate. The impacts should be evaluated to determine those that affect these populations. In considering the impacts to the populations, differential patterns of consumption of natural resources should be considered (i.e., differences in rates and/or pattern of fish, vegetable, water, and/or wildlife consumption among groups defined by demographic factors such as socioeconomic status, race, ethnicity, and/or cultural attributes). The impacts to the local area surrounding the site should be summarized in the environmental justice section of the EIS (or EA if analyzed). It is not necessary to discuss the impacts at the same level of detail as in the impact sections. It is acceptable to briefly mention the impact and reference the section where it is discussed in greater detail.

Next, one should assess if the impacts disproportionately impact the minority or low-income population, i.e., Are the impacts greater for these populations? Are there any impacts experienced by these populations that are not experienced by others? To effectively visualize the impacts, it may be helpful to display the minority and low-income population data spatially. In cases where the population is located next to the site, the impacts or potential for impact will likely be disproportionate for these populations. For instance, potential exposure to effluents may be greater to those living closest to the facility, noise and traffic may disrupt nearby residents to a greater extent than those living far from the site, and the potential risk due to accidents may be greater for nearby residents. If there are no disproportionate impacts, no further analysis would be needed. The reviewer should document the finding in the environmental justice section.

2.   Finally, it is necessary to determine if the impacts are high and adverse. Another way of stating this: Are the impacts significant, unacceptable or above generally accepted norms such as regulatory limits or state and local statutes and ordinances. Each impact, and where appropriate, the cumulative and multiple effect of the impacts, should be reviewed for significance. If the statement can be made that no combination of the impacts are significant, then there are no disproportionate adverse and high impacts on the minority or low-income populations. The reviewer should document the conclusion in the environmental justice section.

D.   If there are significant impacts to the minority or low-income population, it is then necessary to look at mitigative measures and benefits. The reviewer should determine and discuss if there are any mitigative measures that could be taken to reduce the impact. To the extent practicable, mitigation measures should reflect the needs and preferences of the affected minority or low-income populations. The reviewer should discuss the benefits of the project to surrounding communities, even though benefits to a specific group may be difficult to determine, particularly

NOTE: The environmental justice (EJ) guidance contained in this appendix is provided as draft for interim use. The Commission has directed the staff to develop an EJ policy statement. After the policy statement is completed, necessary updates to the EJ guidance will be incorporated.

economic benefits. The conclusion at this point is project specific. The conclusion may be that there are disproportionately high and adverse impacts to minority and low-income populations; however, factors such as the mitigative measures and/or the benefits of a project may outweigh the disproportionate impacts. In any case, the facts should be presented so that the ultimate decision maker can weigh all aspects in making the agency decision. The Executive Order does not prohibit taking an action where there are disproportionate high and adverse impacts to minority and low-income populations.

E.   The results of an environmental justice evaluation should be documented in the EIS or special case EA. The results should indicate if a disproportionately high and adverse human health or environmental impact is likely to result from the proposed action and any alternatives, and should be written in non-technical plain language. The document should contain a distinct section on environmental justice even if the demographics do not indicate a potential for an environmental justice concern. If a site has already received an environmental justice evaluation, it is acceptable to reference the previous evaluation and provide a summary of the findings and then add any new information that results from the proposed action. For instance, if environmental justice is included in a license renewal, it would not need to be completely readdressed for a license amendment.

Following an EIS or EA, the NRC announces its decision in a Record of Decision (ROD) or a FONSI. (For NRC, the ROD is the issuance of the license or license amendment.) For an EIS or special case or circumstance EA, the ROD or FONSI should document the conclusion of the findings on environmental justice, including any mitigative measures that will be taken to reduce the impact.

IV.   POLICY IMPLEMENTATION FOR RULEMAKING ACTIVITIES

A.   The staff responsible for rulemaking should address environmental justice in the preamble to any proposed and final rules that require an EIS, a supplement to an EIS, generic EIS, or if warranted by a special case or circumstance EA/FONSI, as described in Section II, above.

B.   If it is known in advance that a particular rulemaking might impact a specific population disproportionately, the NRC staff should ensure that the population knows about the rulemaking and is given the opportunity to participate. Measures to increase public participation are discussed in Section III. B. above.

C.   If an environmental justice analysis is performed for a rulemaking activity, the staff should include language contained in NUREG/BR-0053, Revision 4, Section 3.13 and 5.13 to the Federal Register Notice to seek and welcome public comments on environmental justice. The staff should follow the "Policy Implementation for Licensing Actions," in Section III above, to perform the environmental justice review.

D.   Public comments on environmental justice issues should be addressed in the preamble to the final rule when published in the *Federal Register*. Environmental justice comments should be addressed at the same level of detail and in the same location as comments received on other parts of the rule.

E.   When a rule is being modified or developed that contains siting evaluation factors or criteria for siting a new facility, the staff may consider including specific language in the rule or supporting regulatory guidance to state that an environmental justice review will be performed as part of the licensing process.

APPENDIX D
CONSULTATION PROCEDURES

State Consultation on Draft Environmental Assessment

The PM should initially contact the appropriate State official by telephone. The licensing PM should:

- Explain the licensing action;
- Inform the State official that NRC has completed a draft EA;
- Describe any impacts from the Environmental Assessment; and
- Ask the State official if they would like to review the EA.

If the State official wishes to review the EA, negotiate a timeframe that does not exceed 30 days.

The licensing PM should transmit a copy of the EA, requesting the State officials views, including the date by which comments must be received. In transmitting the EA, the cover letter should be made publicly available, however the draft EA document **should not** be made publicly available as it is considered pre-decisional information. The EA should include the following text in the header/footer:

> "Pre-decisional Information: This information is limited to use by **[insert State Agencies name]** and their staff. This information shall not be disclosed without prior NRC permission."

If comments are received from the State, the comments should be summarized in the EA. Minor comments could be characterized as "general agreement" or "no objection" by the State or agency, as appropriate. For more extensive comments the licensing PM should summarize the details of the comments and the responses in the EA or place them in a separate document and reference them in the EA. The EA and comment response document should be placed in the NRC Public Document Room to ensure public access.

If the State official does not wish to review the EA or does not provide comments within the agreed upon timeframe, the consultation is complete. The EA should note that the State was consulted and had no comments.

For rulemaking actions, the draft EA is sent to the State Liaison Officer for comment while the proposed rule is out for comment. This is accomplished through the Office of State and Tribal Programs. The "Regulations Handbook" (NUREG-0053) provides additional information.

SECTION 7 CONSULTATIONS WITH THE
U.S. FISH AND WILDLIFE SERVICE

NOTE TO USER: This guidance is intended to provide a general road map to complying with Section 7 of the Endangered Species Act and primarily covers informal consultation. Not all of the steps included here will apply in all situations. If a step does not apply the licensing PM should proceed to the next specified step. Please note that other types of consultation are addressed in the regulations but are not discussed in detail here, since they do not represent typical situations; such as formal, early, special, and emergency consultations and conferences on proposed species. The licensing PM is encouraged to contact EPAB with any questions. Additionally, the FWS regulations for compliance with the Section 7 can be found at 50 CFR 402 and additional information is available on the WWW at found at: <http://endangered.fws.gov/consultations/index.html> including the FWS's *Final ESA Section 7 Consultation Handbook* (March 1998).

It should be noted that even if the proposed action is categorically excluded from NEPA review, the licensing PM should still determine if the proposed action requires consultation under Section 7. If so, the PM should proceed with the appropriate consultation as described below.

Finally, the licensing PM should keep in mind that more may be required at different stages of the consultation process than has been outlined here, depending on the FWS region involved and the nature of the proposed action and the affected environment.

For rulemaking actions, consultations will not typically be required as these types of actions are usually considered administrative in nature, see Step 1 below.

I. Initiate the Section 7 Process

Step 1: Is consultation necessary (i.e., is it an administrative action)?
If the proposed action has the potential to affect the physical environment (i.e., involves construction, demolition, changes to effluents or monitoring, changes in transportation levels, or any other activity affecting physical resources), go to Step 2.

If the proposed action does not have the potential to affect the physical environment (i.e., is an administrative action, such as changing ownership, changing record keeping requirements, or is another action with no potential to affect physical resources), then consultation is not required. The licensing PM should include a justification of why the consultation is not necessary for the proposed action in the Consultation Section of the EA. For example:

> "The NRC staff has determined that Section 7 consultation is not required because the proposed action is administrative/procedural in nature and will not affect listed species or critical habitat."

Step 2: What is the area of the proposed action (i.e., "action area")?
Determine the action area for evaluating direct, indirect and cumulative impacts (see definition in Section VI below). Clearly document the rationale for making the determination. Go to step 3.

II. Informal Consultation with FWS

Informal consultation can include phone calls, meetings, email, conversations, letters, and other informal communication modes. If an EIS is being prepared for the proposed action, the FWS recommends that the informal consultation process be initiated prior to public scoping. If assistance is required in making any of the following determinations, contact an environmental PM in EPAB.

Step 3: Are species/habitat in action area?
Obtain a list of species from the FWS (see example in Appendix E). Review and determine whether listed species or critical habitat could be located within the action area. Choose the applicable scenario:

a.   Species/habitat could be present in the proposed action area. Go to Step 4.

b.   Species/habitat are <u>not</u> present in the proposed action area therefore. Document this finding and the Section 7 consultation is complete. For example:

> "The NRC staff has determined that Section 7 consultation is not required because listed species/habitat are not present in the proposed action area, therefore the proposed action will not affect listed species or critical habitat."

Step 4: Could the proposed action adversely affect species or critical habitat?
An evaluation must be completed to determine if the proposed action could adversely affect a listed species or critical habitat. If an evaluation is submitted by the applicant/licensee, it can be used to make a determination. If an evaluation has not been submitted, contact EPAB for assistance. Once the evaluation is complete, choose the applicable scenario:

a.   The proposed action will not affect listed species or habitat. This finding could be based on a field survey which indicates no listed species or habitat are actually present in the "action area." "No effect" conclusions do not require FWS concurrence. For major construction activities it may be beneficial for the administrative record to request a concurring response from the FWS (see example in Appendix E).

b.   The proposed action may affect, but is not likely to adversely affect listed species or habitat. This is the appropriate conclusion when the effects on listed species or critical habitat are expected to be discountable, insignificant, or completely beneficial (see Section VI). This finding should be forwarded to the FWS for concurrence. Consultation is ended if the FWS concurs (usually in writing, although you could receive an email or document a phone conversation, which should be included in the licensing docket). If the FWS does not concur, go to Step 5.

> NOTE: If the action is a major construction activity and the proposed action may affect listed species or habitat, a biological assessment must be prepared (see Section V for more information about biological assessments). Go to Step 7.

c.   The proposed action may affect only <u>proposed</u> species or <u>proposed</u> habitat. Go to Step 6.

d.   The proposed action is likely to adversely affect listed species or habitat. This is the appropriate conclusion if any adverse effect to listed species or habitat may occur as a direct or indirect result of the proposed action and the effects on listed species or critical habitat are <u>not</u> expected to be discountable, insignificant, or completely beneficial (see Section VI). This finding requires formal consultation with the FWS. Go to Step 7.

These conclusion should be based on either field surveys or other verifiable evidence/documentation and should be clearly documented. The FWS may request the following supporting documentation: biological assessment (if applicable) or other studies, map of the vicinity showing the immediate activity boundary and wider area of impacts, current species/habitat list, and other pertinent information (e.g., construction plans).

Step 5: Mitigation or alternatives possible?
Communicate with the FWS and the applicant/licensee to determine whether any alternative actions or mitigative measures can be implemented. If a satisfactory alternative or mitigative action is identified, and the FWS concurs, consultation is complete. If mitigative actions cannot or will not be undertaken, go to Step 7.

III. Conference on Proposed Species/Habitat

Conference with the FWS is only required if it is determined that the proposed activity may jeopardize the continued existence of a proposed species or adversely modify or destroy proposed critical habitat (see 50 CFR 402.10). See Chapter 6 of the FWS handbook for further information

Step 6: Is conference required?
Based on the information above, determine whether the proposed action could result in the jeopardy of a proposed species and/or destroy or adversely modify proposed critical habitat. If so, initiate the conference process with the FWS (the licensing PM should contact EPAB).

IV. Formal Consultation with FWS

The formal consultation process is described in detail in the FWS handbook, Chapter 4 and at 50 CFR 402.14. When an EIS is concurrently being prepared for a proposed action that requires formal consultation, the formal consultation should be initiated no later than the time of release of the DEIS. Biological assessments must have been completed at the time formal consultation begins.

Step 7: Contact EPAB.
This step assumes that the licensing PM and/or the FWS have determined that the proposed action may adversely affect listed species or their critical habitat. Contact staff in EPAB to discuss initiating formal consultation and conducting biological assessments.

Step 8: Prepare request for formal consultation.
You must initiate formal consultation with a written request to the FWS that includes the following information:

- Description of the proposed action;
- Description of the specific area affected (see Step 3 and definition for action area, below);

- Description of any listed species or critical habitat that may be affected;
- Description of the nature of potential effects on species/habitat, including a cumulative effects analysis;
- Relevant reports (any NEPA documents, biological assessments, other analyses); and
- Other information the FWS may determine necessary.

Note that the FWS requires "the best scientific and commercial data" that is available to you.

Step 9: Submit your request
Send the package to the FWS and a copy to the environmental PM. The FWS has 90 days from the date it receives your package to reach a conclusion and 45 days after that to deliver its conclusion (the biological opinion). The biological opinion will be one of "jeopardy" or "no jeopardy" to listed species and/or of "adverse" or "no adverse modification" of critical habitat.

V. Additional Considerations

Biological assessments:
Biological assessments apply to all actions that are major construction activities, unless the FWS concurs with the NRC's determination of "no effect" on listed species or habitat. If the licensing PM has determined a biological assessment is necessary, contact EPAB. Biological assessments must be submitted to the FWS within 180 days of receipt of a species list from the FWS and *before* construction contracts or related activities are begun. If the species list is over 90 days old at the time the assessment is begun, its current accuracy must be verified. If an EIS is being prepared, biological assessments should be completed prior to the release of the DEIS. Refer to Section 3.4 of the FWS handbook and 50 CFR 402.12 for detailed guidance on biological assessments.

National Marine Fisheries Service:
The NMFS (recently renamed NOAA Fisheries) has joint jurisdiction over Section 7 consultations with the FWS. NMFS consultations take place through the regional offices and are typically conducted when marine life (e.g., whales, sea turtles, salmon, other ocean dwelling species) or habitat may be affected. Some species, however, are under the jurisdiction of both agencies; these cases require that both agencies be involved in the consultation process. Generally, the licensing PM should contact NMFS when the proposed action is in a coastal area.

VI. Definitions

action area - all areas to be affected directly or indirectly by the Federal action and not merely the immediate area involved in the proposed action (50 CFR 402.02). The analysis to determine the extent of the action area must not be limited to either the footprint of the proposed action or to the NRC's area of jurisdiction. It should be a biological determination of the reach of the proposed action on listed species. Subsequent analyses of the environmental baseline, effects of the proposed action, and levels of incidental take are based upon the action area.

beneficial effects - are contemporaneous positive effects without any adverse effects to the species.

discountable effects - are those extremely unlikely to occur.

insignificant effects - relate to the size of the impact and should never reach the scale where a take occurs.

is not likely to adversely affect - the appropriate conclusion when effects on listed species are expected to be discountable, insignificant, or completely beneficial (FWS handbook).

major construction activity - a construction project (or other undertaking having similar physical impacts) which is a major Federal action significantly affecting the quality of the human environment as referred to in NEPA (50 CFR 402.02 or FWS handbook). As a rule of thumb, actions that require EISs and have construction-type impacts (e.g., dams, buildings, pipelines, water resource activities) are typically considered major construction activities.

may affect - conclusion reached when an action may pose *any* effects on species or habitat. When a "may affect" situation exists, either formal consultation is begun or written concurrence with a conclusion of "not likely to adversely affect" is obtained from the FWS (FWS handbook).

SECTION 106 CONSULTATIONS WITH THE
STATE HISTORICAL PRESERVATION OFFICER

NOTE TO USER: This guidance is intended to provide a general road map to complying with Section 106 of the National Historic Preservation Act. All of the steps listed here will not necessarily be required for each individual licensing action. If a step does not apply the licensing PM should proceed to the next specified step. The licensing PM is encouraged to contact EPAB with any questions. Additionally, the Advisory Council on Historic Preservation (Council) regulations for compliance with the Section 106 can be found at 36 CFR 800 and additional information is available on the WWW at <http://www.achp.gov>.

It should be noted that even if the proposed action is categorically excluded from NEPA review, the licensing PM must still determine if the proposed action requires consultation under Section 106. If so, the licensing PM should proceed with the appropriate consultation as described below.

Finally, the licensing PM should keep in mind that more may be required at different stages of the consultation process than has been outlined here, depending on the parties involved and the nature of the proposed action and the affected environment.

For rulemaking actions, consultations will not typically be required as these types of actions are usually considered administrative in nature, see Step 2 below.

I. Initiate the Section 106 Process

Step 1: Does the proposed action qualify as an "undertaking"?
The licensing PM should first determine whether or not an undertaking exists. An undertaking means a project, activity, or program funded in whole or in part under the direct or indirect jurisdiction of NRC, including those carried out by or on behalf of NRC; those carried out with NRC's financial assistance; those requiring a Federal permit, license or approval; and those subject to State or local regulation administered pursuant to a delegation or approval by NRC. If the answer to this question is yes; go to Step 2. NOTE: For NRC licensing actions the answer is always yes.

Step 2: Is the undertaking a type of activity that has the potential to affect historic properties?
Examples of activities with no potential to affect include administrative or procedural license changes (e.g., changes in ownership, other name changes, recordkeeping requirements, etc. that involve no physical changes to the environment or structures). **For the purposes of this determination, it must be assumed that historic properties are present.** If an undertaking does not present a type of activity (i.e., administrative/procedural actions) that has the potential to affect a historic property, then no further consultation under Section 106 is necessary. The licensing PM should include a justification of why the consultation process is not necessary for the proposed action in the Consultation Section of the EA. For example:

> "The NRC staff has determined that the proposed action is not a type of activity that has potential to cause effects on historic properties because it is administrative/procedural action. Therefore, no further consultation is required under Section 106 of the National Historic Preservation Act."

If the undertaking is a type of activity that has the potential to affect historic properties; go to Step 3.

II. Identify Historic Properties

Step 3: Determine and document the "area of potential effects"?
The area of potential effects means the geographic area or areas within which the proposed action (i.e., "undertaking") may directly or indirectly cause alterations in the character or use of historic properties, if any such properties exist. The area of potential effects is influenced by the scale and nature of an undertaking and may be different for different kinds of effects caused by the undertaking. It is not necessarily a contiguous area (for example, several sites where changes may be possible could be the area of potential effects). In documenting this area to the SHPO, the licensing PM may need to submit photographs, maps, or drawings. Go to Step 4.

Step 4: Initiation of consultation.
The licensing PM must send a letter to the SHPO/THPO, and consulting parties if applicable, describing (1) the "undertaking" (i.e., proposed action) and (2) the area of potential effects. The letter should also explain any actions taken by the applicant/licensee to identify historic properties (NOTE: A currently licensed site is likely to have documentation regarding surveys for historic properties).

The letter should also request the SHPO/THPOs views on the scope of identification efforts. These efforts include:

- Determining and documenting the area of potential effects;
- Requesting the views of the SHPO on further actions to identify historic properties that may be affected;
- Seeking information, as appropriate, from consulting parties who may have knowledge are concerns about the area; and
- Gathering information from American Indian tribes and Native Hawaiian organizations about properties which they attach religious and cultural significance.

A map/drawing of the area of potential effect should accompany the letter, if appropriate. For highly complex projects, the licensing PM may wish to schedule meetings or conference calls with the consulting parties. Go to Step 5.

Step 5: Identify Historic Properties.
The standard for identification is a "reasonable and good faith effort" to identify historic properties (e.g, those properties listed in or eligible for the *National Register* and National Historic Landmarks) and depends on a variety of factors, including, but not limited to, previous identification work. The steps taken for identification of a historic property include: preliminary work, actual efforts to identify properties and an evaluation of identified properties. Appropriate identification may include background research, consultation, oral history interviews, sample field investigation, and field surveys. Go to Step 6.

Step 6: Evaluate historic significance.
Based on the information gathered, the licensing PM makes a determination of eligibility by applying the *National Register* criteria to properties identified in the area of potential effects. This determination is

made in consultation with the SHPO. Old determination of eligibility needs to be reevaluated due to passage of time. If no historic properties are found or no historic properties are affected, the licensing PM should notify the consulting parties and the Section 106 process is complete. The licensing PM should include a description of the consultation process for the proposed action in the EA or EIS.

If historic properties may be affected or the SHPO/THPO objects to a no historic properties affected finding; go to Step 7 (NOTE: the Council may need to be contacted as discussed in Step 8).

III. Assess adverse effects

Step 7: Contact EPAB.
This step assumes that the licensing PM and/or the SHPO has determined that historic properties may be affected by the proposed action. Contact EPAB to discuss the following steps and obtaining outside assistance if necessary. Go to Step 8.

Step 8: Apply criteria of adverse effect.
Adverse effects occur when the proposed action may directly or indirectly alter characteristics of a historic property that qualify it for inclusion in the *National Register*. Reasonably foreseeable effects caused by the undertaking that may occur later in time, be farther removed in distance, or be cumulative also need to be considered.

Examples of adverse effects include physical destruction or damage; alteration not consistent with the Secretary of the Interior's Standards; relocation of a property; change of use or physical features of a property's setting; visual, atmospheric, or audible intrusions; neglect resulting in deterioration; or transfer, lease, or sale of a property out of Federal ownership or control without adequate protections.

If a property is restored, rehabilitated, repaired, maintained, stabilized, remediated or otherwise changed in accordance with the Secretary of the Interior Standards, then it will not be considered an adverse effect (assuming that the SHPO agrees). Where properties of religious and cultural significance to American Indian tribes or Native Hawaiian organizations are involved, neglect and deterioration may be recognized as qualities of those properties and thus may not necessarily constitute an adverse effect.

Alteration or destruction of an archaeological site is an adverse effect, whether or not recovery of archaeological data from the site is proposed. The Council has issued guidance to help agencies and others reach agreement on the treatment of such properties.

The consulting parties must be consulted if NRC applies the criteria of adverse effect. If no adverse effects are found, document the findings and provide to the SHPO. If the SHPO fails to respond within 30 days, the licensing PM can consider the SHPO in agreement. No further consultation is required the licensing PM should include a description of the consultation process for the proposed action in the EA or EIS.

If the finding is also submitted to the Council, the licensing PM may proceed with the proposed with the Council's concurrence on the finding or if no response is received within 15 days.

It should be noted that special procedures apply to adverse effects on National Historic Landmarks.

If adverse effects from the proposed action are present, or there is disagreement with the SHPO or Council, go to Step 9.

IV. Resolve adverse effects

Step 9: Continue consultation.
When adverse effects are found, consultation must continue among the consulting parties. The licensing PM should notify the Council when adverse effects are found. The licensing PM should also provide project documentation to all consulting parties to assist in resolving the effects and provide the public with an opportunity to express views.

When resolving adverse effects without the Council, the licensing PM consults with the SHPO and other parties to develop a Memorandum of Agreement (Agreement). If this is achieved, the Agreement is filed with the Council and is the formal conclusion of Section 106. When the Council is involved, the same process occurs, but the Council must also concur on the the the Agreement.

If adverse effects are not resolved with an Agreement, go to Step 10.

V. Failure to resolve adverse effects

Step 10: Proceeding with proposed action.
When consultation is terminated, the Council will render advisory comments to the NRC. This must be considered when the final decision on the undertaking is made.

VI. Additional Considerations

Roles of Participants:
Participants in the Section 106 review process and their roles and responsibilities are outlined in 36 CFR 800.2. Potential participants include:

- Agency Official (800.2(a))
- Council (800.2(b))
- SHPO (800.2(c)(1))
- American Indian tribes when the undertaking occurs or affects historic properties, on their tribal lands (800.2(c)(2)(i))
- American Indian tribes and Native Hawaiian organizations (800.2(c)(2)(ii))
- Representatives of local governments (800.2(c)(3))
- Applicants for Federal assistance, permits, licenses and other approvals (800.2(c)(4))
- Additional consulting parties (800.2(c)(5))
- The public (800.2(d))

In addition, through the identification and evaluation process (800.4), and when National Historic Landmarks may be directly and adversely affected (800.10), the Secretary of the Interior, represented by the National Park Service, may be involved as well.

<u>Involving the Public:</u>
Public involvement is a critical aspect of the Section 106 process. Section 800.2(d) contains a standard that Federal agencies must adhere to as they go through the Section 106 process.

The type of public involvement will depend upon various factors, including but not limited to, the nature and complexity of the undertaking, the potential impact, the historic property, and the likely interest of the public in historic preservation issues.

Section 800.2(d)(2) sets a notice and public information standard. The public must be notified, with sufficient information to allow meaningful comments, so that they can express their views during the various stages and decision-making points of the Section 106 process.

It is intended that Federal agencies have flexibility in how they involve the public, including the use of the NEPA and other agency planning processes, as long as opportunities for such public involvement are adequate and consistent with subpart A of the regulations. Section 800.2(d)(3) provides reminders of this flexibility.

<u>Consultation:</u>
"Consultation" is a dynamic, good-faith process of seeking, discussing, and considering the views of other participants and, where feasible, seeking agreement with them regarding matters arising in the Section 106 process. The Section 110 standards and guidelines issued by the Secretary of the Interior provide additional guidance on consultation as part of Standard 5: "An agency consults with knowledgeable and concerned parties outside the agency about its historic preservation-related activities."

Throughout the Section 106 process, the Agency Official (i.e., the Division Director responsible for the approving the proposed action) is to involve the consulting parties described in Section 800.2(c) in findings and determinations made during the Section 106 process. Additional guidance for Federal agencies on consultation is found in Section 800.2(a)(4).

During resolution of adverse effects (Section 800.6), the purpose of consultation is to seek ways to avoid, minimize, or mitigate any adverse effects, and to incorporate the results of that consultation into a Memorandum of Agreement.

<u>Seeking Information:</u>
The licensing PM should involve the consulting parties in findings and determinations made during the Section 106 process. The SHPO reflects the interests of the State and its citizens in the preservation of their cultural heritage and must be consulted in making determinations. For a tribe that has assumed the responsibilities of the SHPO for section 106 on tribal lands under section 101(d)(2) of the act, the Tribal Historic Preservation Officer (THPO) appointed or designated in accordance with the act is the official representative for the purposes of Section106. Tribes that have not been certified still have the same consultation and concurrence rights as THPOs when the undertaking takes place, or affects historic properties, on their tribal lands. The practical difference is that during undertakings with THPOs, the THPO would be consulted in lieu of the SHPO, while non-certified tribes would be consulted in addition to the SHPO. The licensing PM should identify and consult with any American Indian tribe or Native Hawaiian organization (not limited to activities in Hawaii) that attaches religious and cultural significane to historic properties that may be affected by the undertaking. The applicant/licensee and any local

governments with jurisdiction over the area of potential effects is entitled to participate as a consulting parties. Certain individuals and organizations with a demonstrated interest in the undertaking may participate as consulting parties due to the nature of their legal or economic relation to the undertaking or affected properties, or their concern with the undertaking's effects on historic properties. In addition, the licensing PM should contact the SHPO/THPO to identify any other parties entitled to be consulting parties (see 36 CFR 800.2(c) for more detail).

Documentation:
The Agency Official is supposed to ensure that a determination, finding, or agreement reached through the Section 106 process is supported by sufficient documentation to enable any reviewing parties (including the public) to understand its basis.

Section 800.11 contains details of documentation standards, as well as general guidance on the adequacy of documentation, its format, and issues of confidentiality.

VII. Definitions

The following is a partial list of definitions as provided in 36 CFR 800:

Act - means the National Historic Preservation Act of 1966, as amended, 16 U.S.C. 470-470w-6.

Historic property - means any prehistoric or historic district, site, building, structure, or object included in, or eligible for inclusion in, the National Register of Historic Places maintained by the Secretary of the Interior. This term includes artifacts, records, and remains that are related to and located within such properties. The term includes properties of traditional religious and cultural importance to an American Indian tribe or Native Hawaiian organization and that meet the National Register criteria.

eligible for inclusion in the National Register - includes both properties formally determined as such in accordance with regulations of the Secretary of the Interior and all other properties that meet the National Register criteria.

National Historic Landmark - means a historic property that the Secretary of the Interior has designated a National Historic Landmark.

National Register - means the National Register of Historic Places maintained by the Secretary of the Interior.

National Register criteria - means the criteria established by the Secretary of the Interior for use in evaluating the eligibility of properties for the National Register (36 CFR 60).

State Historic Preservation Officer (SHPO) - means the official appointed or designated pursuant to section 101(b)(1) of the act to administer the State historic preservation program or a representative designated to act for the State historic preservation officer.

Tribal Historic Preservation Officer (THPO) - means the tribal official appointed by the tribe's chief governing authority or designated by a tribal ordinance or preservation program who has assumed the responsibilities of the SHPO for purposes of section 106 compliance on tribal lands in accordance with section 101(d)(2) of the act.

Undertaking - means a project, activity, or program funded in whole or in part under the direct or indirect jurisdiction of a Federal agency, including those carried out by or on behalf of a Federal agency; those carried out with Federal financial assistance; those requiring a Federal permit, license or approval; and those subject to State or local regulation administered pursuant to a delegation or approval by a Federal agency.

PAGE INTENTIONALLY BLANK

APPENDIX E
EXAMPLE LETTERS AND DOCUMENTS

CONSULTATION LETTERS

ENVIRONMENTAL ASSESSMENT "COMPLEX"

ENVIRONMENTAL ASSESSMENT "SIMPLE"

ENDANGERED SPECIES CONSULTATION LETTERS:
"EXAMPLE REQUEST FOR SPECIES LIST"

Name, Title
U.S. Fish and Wildlife Service
Address

SUBJECT: REQUEST FOR INFORMATION REGARDING ENDANGERED SPECIES AND CRITICAL HABITAT FOR THE PROPOSED **[INSERT PROJECT NAME OR TITLE]**

Dear _____,

[Provide brief description of the proposed action.]

[Provide a description of the "action area." This may include maps, pictures, coordinates, as necessary.] Please provide information that you may have regarding the presence of endangered or threatened species or critical habitat in the action area.

After assessing the information provided by you, we will determine what additional actions are necessary to comply with the Section 7 consultation process.

If you have any questions, please contact _____.

Sincerely,

Name, Title
Branch
Division
Office

Docket No./License No:

ENDANGERED SPECIES CONSULTATION LETTER:
"EXAMPLE REQUEST FOR CONCURRENCE ON DETERMINATION OF EFFECT ON ENDANGERED SPECIES"

Name, Title
U.S. Fish and Wildlife Service
Address

SUBJECT: REQUEST FOR CONCURRENCE ON THE DETERMINATION OF EFFECTS ON FEDERALLY LISTED SPECIES AND THEIR CRITICAL HABITATS FOR THE **[INSERT PROJECT NAME OR TITLE]**

Dear _____,

[Provide brief background. Include reference to previous correspondence or telephone conversations.]

After a review of the potential impacts of the proposed action, we have determined that the proposed action "may affect" listed species or their designated critical habitat, however, these effects are expected to be **[Choose one of the following:]** discountable **[OR]** insignificant, **[see definition in Appendix C and provide basis for finding]**, and therefore, have concluded that the proposed action is "not likely to adversely affect" any endangered or threatened species or critical habitat within the area of influence for the proposed action. The supporting basis for this conclusion is enclosed **[Provide documentation, e.g., surveys, analysis, EA, EIS, as applicable]**. We request your concurrence with NRC's determination of "not likely to adversely affect" any listed species or their critical habitat.

If you have any questions, please contact _____.

Sincerely,

Name, Title
Branch
Division
Office

Docket No./License No:

ENDANGERED SPECIES CONSULTATION LETTER: "EXAMPLE REQUEST FOR CONCURRENCE ON DETERMINATION OF NO EFFECT FOR MAJOR CONSTRUCTION ACTIVITIES"

Name, Title
U.S. Fish and Wildlife Service
Address

SUBJECT: REQUEST FOR CONCURRENCE ON NRC'S DETERMINATION OF NO EFFECTS ON FEDERALLY LISTED SPECIES AND THEIR CRITICAL HABITATS FOR THE **[INSERT PROJECT NAME OR TITLE]**

Dear _____,

[Provide brief background, including the "major construction activity." Include reference to previous correspondence or telephone conversations.]

After a review of the potential impacts of the proposed action, we have determined that the proposed action will have no effect on any endangered or threatened species or critical habitat within the area of influence for the proposed action. The supporting basis for this conclusion is enclosed **[Provide documentation, e.g., surveys, analysis, EA, EIS as applicable]**. We request your concurrence with NRC's determination of "no effect" on any listed species or their critical habitat.

If you have any questions, please contact _____.

Sincerely,

Name, Title
Branch
Division
Office

Docket No./License No:

SHPO CONSULTATION LETTER:
"INITIATION OF SECTION 106 CONSULTATION"

Name
State Historical Preservation Officer
Address

SUBJECT: INITIATION OF SECTION 106 PROCESS FOR **[INSERT PROJECT NAME OR TITLE]**

Dear _____ ,

[Provide brief background of the licensing action. Include reference to previous correspondence or telephone conversations, relevant documentation.]

[Provide a description of the "area of potential affects." This may include maps, pictures, coordinates, as necessary.]

As required by 36 CFR 800.4(a), the NRC is requesting the views of the State Historical Preservation Officer on further actions to identify historic properties that may be affected by the NRC's proposed action (i.e., **[licensing the proposed _____ or granting the proposed licensed amendment to _____ Facility]**). **[Describe any additional information you will use such as surveys, interviews etc., if appropriate]** and any information you provide will be used to document affects in accordance with 36 CFR 800.4(d).

After assessing the information provided by you, we will determine what additional actions are necessary to comply with the Section 106 consultation process.

If you have any questions, please contact _____ .

Sincerely,

Name, Title
Branch
Division
Office

Docket No./License No:

SHPO CONSULTATION LETTER:
"EXAMPLE REQUEST FOR CONCURRENCE ON DETERMINATION OF EFFECT FOR HISTORIC PROPERTIES"

Name
State Historical Preservation Officer
Address

SUBJECT: REQUEST FOR CONCURRENCE ON THE DETERMINATION OF EFFECTS ON HISTORIC PROPERTIES FOR THE **[INSERT PROJECT NAME OR TITLE]**

Dear _____,

[Provide brief background. Include reference to previous correspondence or telephone conversations, relevant documentation.]

After a review of the potential impacts of the proposed action, we have determined that **[Choose one of the following:]** there are no historical properties within the "area of potential effect" **[OR]** there are historical properties present, however we have concluded that the proposed action will not adversely affect these historic properties. The supporting basis for this conclusion is enclosed **[Provide documentation, e.g., surveys, analysis, EA, EIS, as applicable]**. We request your concurrence with this determination that the proposed action does not adversely affect any historical properties.

If you have any questions, please contact _____.

Sincerely,

Name, Title
Branch
Division
Office

Docket No./License No:

ENVIRONMENTAL ASSESSMENT
"COMPLEX LICENSING ACTION"

Due to the length of these example documents, the reader is referred to ADAMS to view example Environmental Assessment's for complex licensing actions. NOTE: the term "complex" refers solely to the EA format as described in Section 3.4 of this guidance.

Environmental Assessment Related to Approval of the Maine Yankee License Termination Plan, ADAMS Accession No.: ML030340122.

Environmental Assessment Related to Nuclear Fuel Services License Amendment Request to Construct and Operate Uranyl Nitrate Storage Building, ADAMS Accession No.: ML021790068.

Environmental Assessment related to the International Uranium Corporation's White Mesa Uranium Mill Amendment Request for the Receipt and Processing of the Maywood Alternate Feed, ADAMS Accession No.: ML022350454.

ENVIRONMENTAL ASSESSMENT
"SIMPLE LICENSING ACTION"

MEMO TO: Docket File **[Insert docket/license number]**

FROM: **Name, Title**
Branch
Division, Office

SUBJECT: ENVIRONMENTAL ASSESSMENT FOR **[Insert Title of Proposed Project, County, State]**, IN CONSIDERATION OF AN AMENDMENT TO LICENSE **[Insert License No.]** FOR **[Describe what the license would allow.]**

This Environmental Assessment (EA) is being performed to evaluate the environmental impacts of the proposed amendment to **[applicant/licensee name, license number, and facility, e.g., ABC's Company's License 123, at Anytown, USA]**. **[Applicant/licensee name]** submitted a license application **[OR]** amendment application by letter dated **[insert date]** to amend its license to allow **[brief description of what the amendment would allow]**.

Based upon the analysis contained in this EA, the staff concludes that the proposed action will not have a significant effect on the quality of the human environment. Accordingly, the staff has determined that preparation of an environmental impact statement is not warranted. This conclusion will be documented in the *Federal Register* as required by 10 CFR 51.35

License No.: _____
Attachment: Environmental Assessment

U.S. NUCLEAR REGULATORY COMMISSION
DOCKET NO. **XXXX**
[Date]
Environmental Assessment Related to Issuance of a License Amendment
of U.S. Nuclear Regulatory Commission **XXXX** Materials License No. **XXXXXX,**
[name of licensee] in **City, State**

Introduction:

The U.S. Nuclear Regulatory Commission (NRC) staff has prepared this environment assessment of the ABC Corporation's (ABC's or the licensee's) decommissioning plan for its Anytown site. The XYZ facility is operated by ABC in Anytown, State. ABC was authorized by NRC from 1973 to 1998 to use radioactive materials for nuclear medicine purposes at the site. In 1998, ABC ceased operations at the XYZ facility and requested that NRC terminate its license. ABC has conducted characterization surveys of the facilities and identified carbon-14 (C-14) and tritium (H-3) contamination in the XYZ nuclear medicine facilities. The NRC staff has evaluated ABC's request and has developed an environmental assessment (EA) to support the review of ABC's proposed decommissioning plan and license amendment request, in accordance with the requirements of 10 CFR Part 51. Based on the staff evaluation, the conclusion of the EA is a Finding of No Significant Impact (FONSI) on human health and the environment for the proposed licensing action.

[Describe the proposal. Briefly characterize the location and contamination and reference the decommissioning plan or license termination request.] The XYZ facility incorporates 10 buildings on 40 acres located at 123 East Main Street in Anytown. ABC conducted a characterization survey of the affected areas and developed a decommissioning plan. The survey confirmed the presence of H-3 contamination in portions of the facility and was used as the basis for development of the decommissioning plan. The affected area of the XYZ facility consists of the former nuclear medicine laboratory and associated rooms in the basement of one building, identified as Building One. ABC proposed to use the screening values developed by NRC **(see NUREG-1757, Vol. 1, Appendix B)** as the derived concentration guideline levels (DCGLs) for decommissioning and as the basis for demonstrating that the site meets NRC's radiological cleanup criteria.

The Proposed Action:

[Describe the proposal. Briefly summarize the remediation activities and reference the decommissioning plan or license termination request for a more thorough description.] The proposed action is to amend NRC Radioactive Materials License Number 31-XXXX to incorporate appropriate and acceptable DCGLs into the license. The licensee's objective for the decommissioning project, as stated in the decommissioning plan, is to decontaminate and remediate the affected areas of Building One sufficiently to enable unrestricted use, while ensuring exposures to occupational workers and the public during the decommissioning are maintained as low as reasonably achievable (ALARA). ABC's decommissioning plan for the XYZ facility proposes to use DCGLs that are screening values developed by NRC (65 FR 37186, June 13, 2000) to demonstrate compliance with the radiological criteria for license termination in 10 CFR Part 20.1402. The DCGLs will define the maximum amount of residual radioactivity on building surfaces, equipment and materials and in soils, that will satisfy the NRC requirements in Subpart E, 10 CFR Part 20, "Radiological Criteria for License Termination." The DCGLs proposed to be incorporated into the license are as follows:

Derived Concentration Guideline Levels			
Radionuclide	Release of equipment & materials (surfaces)	Building surfaces	Soil
H-3			
C-14			

Need for the Proposed Action:

The purpose of the proposed action is to reduce residual radioactivity at the XYZ facility to a level that permits release of the property for unrestricted use and termination of the license. NRC is fulfilling its responsibilities under the Atomic Energy Act to make a decision on a proposed license amendment for decommissioning that ensures protection of the public health and safety and environment.

The Environmental Impacts of the Proposed Action:

[Briefly summarize special environmental or cultural issues that may be associated with a decommissioning action and may require a particular analysis. Include radiological and nonradiological direct and indirect impacts - including: ecological; aesthetic; historical; cultural; socioeconomic; and health. Also, include a paragraph on adverse impacts, cumulative impacts and the evaluation of the significance of the impacts]

The NRC staff has reviewed the decommissioning plan for the XYZ facility and examined the impacts of decommissioning. Based on its review, the staff has determined that the affected environment and the environmental impacts associated with the decommissioning of the XYZ facility are bounded by the impacts evaluated by the "Generic Environmental Impact Statement in Support of Rulemaking on Radiological Criteria for License Termination of NRC-Licensed Nuclear Facilities" (NUREG-1496). The staff also finds that the proposed decommissioning of the XYZ facility is in compliance with 10 CFR Part 20.1402, the radiological criteria for unrestricted use.

Since ceasing operations, the XYZ site has been stabilized to prevent contamination from spreading beyond its current locations. Access to the contaminated areas is controlled to assure the health and safety of workers and the public. No ongoing licensed activities are occurring in the facilities.

Contamination controls will be implemented during decommissioning to prevent airborne and surface contamination from escaping the remediation work areas, and therefore no release of airborne contamination is anticipated. However, the potential will exist for generating airborne radioactive material during decontamination, removal and handling of contaminated materials. If produced, any effluent from the proposed decommissioning activities will be limited in accordance with NRC requirements in 10 CFR Part 20 or contained onsite or treated to reduce contamination to acceptable levels before release, and shall be maintained ALARA. Release of contaminated liquid effluents are not expected to occur during the work.

ABC and subcontractors will perform the remediation under the XYZ license, with ABC overseeing the activities and maintaining primary responsibility. The XYZ facility has adequate radiation protection procedures and capabilities, and will implement an acceptable program to keep exposure to radioactive materials ALARA. As noted above, ABC has prepared a decommissioning plan describing the work to be performed, and work activities are not anticipated to result in a dose to workers or the public in excess of the 10 CFR Part 20 limits. Past experiences with decommissioning activities at sites similar to the XYZ facility indicate that public and worker exposure will be far below the limits found in 10 CFR Part 20.

Environmental Impacts of the Alternatives to the Proposed Action

[Describe reasonable alternatives. A no-action alternative should always be considered.] The only alternative to the proposed action of allowing decommissioning of the site is no action. The no-action alternative is not acceptable because it will result in violation of NRC's Timeliness Rule (10 CFR Part 30.36), which requires licensees to decommission their facilities when licensed activities cease, and to request termination of their radioactive materials license.

Agencies and Persons Consulted:

This EA was prepared by NRC staff **[list staff and title]** and coordinated with the following agencies: State Department of Environmental Quality, State Office of Historical Preservation, State Fish and Wildlife Service, and the U.S. Fish and Wildlife Service. NRC staff provided a draft of its Environmental Assessment to [**State agency**] for review.

On [**provide date**], the [**State agency**] responded by [**telephone, letter, etc**] and stated that it had no comments [**Or explain comments.**]

Conclusion:

The NRC staff has concluded that the proposed action complies with 10 CFR Part 20. Decommissioning of the site to the DCGLs proposed for this action will result in reduced residual contamination levels in the facility, enabling release of the facility for unrestricted use and termination of the radioactive materials license. No radiologically contaminated effluents are expected during the decommissioning. Occupational doses to decommissioning workers are expected to be low and well within the limits of 10 CFR Part 20. No radiation exposure to any member of the public is expected, and public exposure will therefore also be less than the applicable public exposure limits of 10 CFR Part 20. NRC has prepared this EA in support of the proposed license amendment to incorporate appropriate and acceptable DCGLs and to use the proposed DCGLs for the planned decommissioning by the licensee at the XYZ facility. Based upon the analysis contained in this EA, the NRC staff concludes that the proposed action will not have a significant effect on the quality of the human and has determined not to prepare an environmental impact statement for the proposed action.

Sources Used:

The following references are available for inspection at NRC's Public Electronic Reading Room at <http://www.nrc.gov/NRC/ADAMS/index.html>.

[Reference documents used in preparing EA using a complete reference format, i.e., author, title, date, and ADAMS reference numbers.]

APPENDIX F
GLOSSARY

GLOSSARY

aquifer—A body of rock or soil that can conduct groundwater and can yield significant quantities of groundwater to wells and springs.

aquifer system—A heterogeneous body of interbedded permeable and poorly permeable material that functions regionally as a water-yielding unit; it comprises two or more permeable beds separated at least locally by confining beds that impede groundwater movement but do not greatly affect the regional hydraulic continuity of the system; includes both saturated and unsaturated parts of permeable material.

archaeological—the study of the buildings, graves, tools and other objects which belonged to people who lived in the past, in order to learn about their culture and society.

ALARA—acronym for "as low as is reasonably achievable" means making every reasonable effort to maintain exposures to radiation as far below the dose limits in 10 CFR Part 20 as is practical consistent with the purpose for which the licensed activity is undertaken, taking into account the state of technology, the economics of improvements in relation to state of technology, the economics of improvements in relation to benefits to the public health and safety, and other societal and socioeconomic considerations, and in relation to utilization of nuclear energy and licensed materials in the public interest (10 CFR 20.1003).

atmospheric diffusion—Refers to the dilution of pollutants in the atmosphere. Diffusion is largely a result of turbulence in the atmosphere and is dependent on the variability characteristics of the wind at the site.

atmospheric dispersion and transport—Refers to the movement or transport of pollutants horizontally or vertically by the wind.

atmospheric stability—An expression of the resistance of the atmosphere to vertical air motion, or dispersion. Stable air resists movement of air upward; unstable conditions result in good vertical dispersion. Atmospheric stability is important to the dispersion and dilution of air contaminants.

atmospheric stagnation—Persistent atmospheric conditions with limited vertical and horizontal air motion, resulting in an increase in the concentration of air contaminants.

benthic—Referring to bottom-dwelling aquatic organisms.

biota—The flora and fauna of an area.

climatology—The scientific study of climates (manifestation of weather) over long periods of time.

cofferdam—A watertight enclosure from which water is pumped to expose the bottom of a body of water and permit construction (as of a pier).

consumptive use—The total water loss from a water supply or system by evaporation or transpiration from a vegetated or non-vegetated surface, commercial or industrial process, and all domestic and municipal uses. The difference between the quantity of water withdrawn from a source and the quantity returned to the source or another source of usable water.

cumulative impact—The impact on the environment that results from the incremental impact(s) of an action when added to other past, present, and reasonably foreseeable future actions. Cumulative impacts can result from individually minor but collectively significant actions taking place over a period of time.

dBA (decibel, A-weighted)—A measurement of sound approximately the sensitivity of the human ear and used to characterize the intensity or loudness of sound.

decommissioning—The process of safely removing from service a facility in which nuclear materials are handled, reducing residual radioactivity to a level that permits release of the property for unrestricted use and termination of the NRC license or restricted release (10 CFR 20.1003).

defoliant—A herbicide designed to remove leaves from trees and shrubs or to kill plants.

density-induced current—A gravity-induced flow of one current through, over, or under another, owing to density differences. Factors affecting density differences include temperature, salinity, and concentration of suspended particles.

deposition—Material that is deposited; a deposit or sediment. The laying, placing, or throwing down of any material.

design basis flood—A flood event that is the largest flood against which the various components of a system or facility is protected.

dewatering—Physical removal of water by damming, pumping, diverting, etc.

dewpoint temperature—The temperature at which air becomes saturated with water vapor on being cooled.

dike—An embankment or ridge of either natural or man-made materials used to prevent the movement of liquids, sludges, solids, or other materials.

dredging—Scooping up or excavating (or removing) of earth material or sediment from the bottom of a body of water, raising it to the surface.

ecological—of or pertaining to the environment as it relates to living organisms.

effects—Include: (i) direct effect, which are caused by the action and occur at the same time and place; (ii) indirect effects caused by the action and are later in time or farther removed in distance, but still reasonably foreseeable; and (iii) cumulative effects caused by the aggregate effects of past, present, and reasonably foreseeable future actions. Effects and impacts as used in thi document are synonymous.

effluent—A liquid discharged as waste, such as contaminated water from a facility.

emission—Gases, particles, or liquids released into the atmosphere from smokestacks, other vents, and surface areas of commercial or industrial facilities.

endangered species—As defined in the Endangered Species Act (16 U.S.C. Section 1532, *Definitions*), an endangered species means any species that is in danger of extinction throughout all or a significant portion of its range other than a species of the Class Insecta determined by the Secretary to constitute a pest whose protection under the provisions of this chapter would present an overwhelming and overriding risk to man.

environmental impact statement—A detailed written statement as required by Section 102(2)(C) of the National Environmental Policy Act.

environmental project manager—The project manager who is responsible for reviewing EAs and the environmental review associated with preparation of an EIS. In this document, NUREG-1748, the term is used to generically describe NRC staff in the Environmental and Performance Assessment Branch.

environmental justice—Identifying and addressing, as appropriate, disproportionately high and adverse human health or environmental effects of programs, policies, and activities on minority populations and low-income populations.

environmental monitoring—The process of sampling and analyzing environmental media in and around a facility to: (i) confirm compliance with performance objectives; and (ii) detect contamination entering the environment to facilitate timely remedial action.

environmental review—The process in which the NRC looks at environmental impacts/concerns and documents this review in a NEPA document.

erosion—The wearing away of land surface by wind or water, intensified by land-clearing practices related to farming, residential or industrial development, road building, or logging.

estuary—Region of interaction between rivers and near-shore ocean waters, where tidal action and river flow create a mixing of fresh and salt water. These areas may include bays, mouths of rivers, salt marshes, and lagoons. These brackish water ecosystems shelter and feed marine life, birds, and wildlife.

facility—The building, structure and all components associated with the applicant/licensee for conducting business. May also include other facilities, as necessary, for cumulative impact assessment.

flood—An event in which a river, lake, ocean, or other water feature to rise above normal limits.

floodplain—The lowland and relatively flat areas adjoining creeks, rivers, lakes, and coastal waters. This includes, at a minimum, that area subject to a 1 percent or greater chance of flooding in any given year. The base floodplain shall be used to designate the 100-yr floodplain (1-percent chance floodplain).

fugitive dust—Particulate matter composed of soil surface; can include emissions from haul roads, wind erosion of exposed soil surfaces, and other activities in which soil is removed or redistributed.

geologic—Of or related to a natural process acting as a dynamic physical force on the Earth (faulting, erosion, mountain building resulting in rock formations, etc.).

geologic repository—A system for disposing radioactive waste in excavated geologic media.

geotechnical—Related to geotechnics, a term currently employed to cover the fields of soil mechanics, rock mechanics, engineering geology, and ground improvement.

groundwater—Water contained in pores or fractures in either the *unsaturated zone* or *saturated zone* below ground level.

habitat—Area in which a plant or animal lives and reproduces.

hazardous material—A substance or material in a quantity and form which may pose an unreasonable risk to health and safety or property when transported in commerce.

hazardous waste—As defined in RCRA (42 U.S.C. Section 6903, *Definitions*), a hazardous waste is a solid waste, or a combination of solid wastes, that because of its quantity, concentration, or physical, chemical, or infectious characteristics may: (i) cause, or significantly contribute to an increase in mortality or an increase in serious irreversible, or incapacitating reversible, illness; or (ii) pose a substantial present or potential hazard to human health or the environment when improperly treated, stored, transported, or disposed of, or otherwise managed (40 CFR 261.3).

herbicide—A chemical agent (often synthetic) capable of killing or causing damage to certain plants (usually weeds) without significant disruption of other plants.

historic property—Any prehistoric or historic district, site, building, structure, or object included in, or eligible for inclusion in, the National Register of Historic Places maintained by the Secretary of the Interior. This term includes artifacts, records, and remains that are related to and located within such properties. The term includes properties of traditional religious and cultural importance to an American Indian tribe or Native Hawaiian organization and that meet the National Register criteria [36 CFR 800.16(l)].

hydraulic gradient—Refers to the flow of groundwater. Groundwater flows from areas of higher energy (or hydraulic *head*) to areas of lower hydraulic *head*. The change in hydraulic *head* per unit distance is the hydraulic gradient. Groundwater (and any contaminants moving with it) will flow from upgradient areas to downgradient areas. These terms are analogous to "upstream" and "downstream" flow of surface water.

hydrology—(i) The study of water characteristics, especially the movement of water. (ii) The study of water, involving aspects of geology, oceanography, and meteorology.

impact—The positive or negative effect of an action (past, present, or future) on the natural environment (land use, air quality, water resources, geological resources, ecological resources, aesthetic and scenic resources) and the human environment (infrastructure, economics, social, and cultural). (See *Effects*.)

impoundment—A body of water confined by a dam, dike, floodgate, or other barrier.

inversion—The condition in which air temperature increases with increasing altitude over a certain altitude range. The inversion layer can be at ground level or aloft. The condition results in a layer of warmer air above cooler air, a circumstance that inhibits atmospheric mixing and dispersion of pollutants.

L_{dn} *(day-night sound level)*—The 24-hr time of day weighted equivalent sound level, in decibels, for any continuous 24-hr period, obtained after addition of ten decibels to sound levels produced in the hours from 10p.m. to 7a.m.

L_{eq} *(equivalent sound level)*—The equivalent sound level, in decibels of the mean-square A-weighted sound pressure during a stated time period, with reference to the square of the standard reference sound pressure of 20 micropascals. It is the level of the sound exposure divided by the time period.

licensing project manager—The NMSS project manager who has overall responsibility for the proposed action. In this document, NUREG-1748, the term is used to generically describe NRC staff responsible for determining if a CATX applies and preparing EAs.

levee—(i) A ridge or embankment of sand and silt, built by a stream on its flood plain along both banks of its channel. (ii) An artificial embankment built along the bank of a watercourse or an arm of the sea, to protect land from inundation or to confine streamflow to its channel.

meteorology—The study of the atmosphere and weather conditions in the atmosphere. Knowledge of this science is required for an understanding of the movement and activities of pollutants released into the atmosphere.

mitigation—Actions and decisions that: (i) avoid impacts altogether by not taking a certain action or parts of an action; (ii) minimize impacts by limiting the degree or magnitude of an action; (iii) rectify the impact by repairing, rehabilitating, or restoring the affected environment; (iv) reduce or eliminate the impact over time by preservation and maintenance operations during the life of the action; or (v) compensate for an impact by replacing or providing substitute resources or environments.

mixed waste—a type of waste that contains both hazardous and radioactive source, special nuclear, or byproduct material as defined by the Atomic Energy Act of 1954.

model—A simplified representation of an object or natural phenomenon. The model can be in many possible forms: a set of equations or a physical, miniature version of an object or system constructed to allow estimates of the behavior of the actual object or phenomenon when the values of certain variables are changed. Important environmental models include those estimating the transport, dispersion, and fate of chemicals in the environment.

mrem/yr (millirem per year)—One one-thousandth (0.001) of a *rem* per year.

mSv/yr—One one-thousandth (0.001) of a *sievert* per year.

non-attainment area—An area in which the maximum allowable air concentration of a given pollutant has been reached, therefore additional releases of the pollutant are not permitted.

permeability—The capacity of such media as rock, sediment, and soil to transmit liquid or gas. Permeability depends on the substance transmitted (oil, air, water, etc.) and on the size and shape of the pores, joints, and fractures in the medium and the manner in which they interconnect. "Hydraulic conductivity" means "permeability" in technical discussions relating to *groundwater*.

pH—A number indicating the acidity or alkalinity of a solution. A pH of 7 indicates a neutral solution. Lower pH values indicate more acidic solutions while higher pH values indicate alkaline solutions.

physiochemical—Being physical and chemical. Of or relating to chemistry that deals with the physiochemical properties of substances.

piezometric level—See *potentiometric surface*.

potentiometric surface—(i) The level to which water in a confined aquifer will rise under its own pressure in a borehole, the confining pressure having been removed. (ii) The level to which water will rise in a piezometer (a tube inserted into the ground so that the lower end, with a permeable tip, is at a position from which pre-pressure measurements are required).

proposed action—Action under consideration.

radiation—Refers to the process of emitting energy in the form of rays or particles that are released by disintegrating atoms. NRC is responsible for regulating and licensing the use and possession of certain radioactive materials and the resulting radiation from byproduct and special nuclear materials.

radioactive waste—Solid, liquid, and gaseous materials from nuclear operations that are radioactive or become radioactive and for which there is no further use.

radioactivity—A property possessed by some elements, such as uranium, whereby alpha, beta, or gamma rays are spontaneously emitted.

radionuclide—A radioactive atomic nuclide, which is an atomic nucleus specified by atomic weight, atomic number, and energy state.

reasonable alternatives—Those alternatives that are practical or feasible from the technical and economic standpoint and using common sense.

region (socioeconomic)—The relevant region is limited to that area necessary to include social and economic base data for: (i) the county in which the proposed facility would be located; and (ii) those specific portions of surrounding counties and urbanized areas from which the construction/refurbishment work force would be principally drawn, or that would receive stresses to community services by a change of residence of construction/refurbishment/decommissioning workers. Other social and economic impacts can generally be presumed to fall within the same area covered by this definition of the region.

rem—Rem is the special unit of any of the quantities expressed as dose equivalent equal to the absorbed dose in rads (rad is the special unit of absorbed dose) multiplied by the quality factor (10 CFR 20.1004).

reservoir—A natural or artificial lake used for the storage of water for industrial and domestic purposes and for the regulation of inland waterway levels. Service reservoirs store water for domestic supply purposes under cover and regulate diurnal fluctuations in demand. Impounding reservoirs provide storage to cover seasonal or year-to-year variations in inflow. Such reservoirs (feeder reservoirs) may supply water for domestic or industrial use or for regulating water levels in rivers and canals.

saturated zone—The portion of the ground wholly saturated with water.

scope—Consists of the range of actions, alternatives, and impacts to be considered in an EIS.

scoping—An open process for determining the scope of issues to be addressed and for identifying issues related to a proposed action.

sediment—(i) Solid material that has settled down from a state of suspension in a liquid. (ii) Solid fragmental material transported and deposited by wind, water, or ice, chemically precipitated from solution, or secreted by organisms, and that forms in layers in loose unconsolidated form (e.g., sand, mud, till).

seepage—Percolation of water through the soil from unlined canals, ditches, laterals, watercourses, or water storage facilities.

seismicity—A seismic event or activity such as an earthquake or earth tremor; seismic action.

sewage—Wastewater from homes, businesses, or industries.

site—The area of land owned or controlled by the applicant for the principal purpose of constructing and operating a facility. As a general rule, the applicant's "site boundary" should be accepted as defining the site.

sievert (Sv)—The SI unit of radiation dose equivalent, equal to 1 joule of energy per kilogram of absorbing tissue. The sievert replaces the *rem* (1 Sv = 100 rem).

significantly—See definition in 40 CFR 1508.27.

spoil—(i) The refuse or rubble that accumulates when soil, rock or sand is removed to allow access to mineral deposits. (ii) The material that is removed from a channel when it is dredged.

sole-source aquifer—An aquifer that supplies 50 percent or more of the drinking water of an area.

spawning area—An area used by species for reproduction or deposition of its offspring.

specific—The amount of water a unit volume of saturated permeable rock will yield when drained by gravity.

storage coefficient—The volume of water an aquifer releases from or takes into storage, per unit surface area of the aquifer per unit change in head.

stratification—The arrangement of the waters of a lake in layers of differing density.

stratigraphy—The study of the formation, composition, and sequence of sediments, whether consolidated or not.

stressor—An agent, condition, or other stimulus that causes stress to an organism or other system.

subsidence—The sudden sinking or gradual downward settling of the Earth's surface with little or no horizontal motion.

suspended load—The part of the total stream load that is carried for a considerable period of time in suspension, free from contact with the stream bed; it consists mainly of clay, silt, and sand.

tectonic—Of, or relating to, tectonics.

tectonics—Geological structural features as a whole. A branch of geology concerned with the structure of the crust of a planet or moon and especially with the formation of folds and faults in it.

unsaturated zone—The area between the surface and the upper limit of the saturated zone (water table) where only some of the spaces (fractures and rock pores) are filled with water.

vector—An organism, often an insect or rodent, that carries disease.

vicinity—The surrounding area of the proposed action. Depending on the action and environmental media being considered this can range from less than one mile to 50 miles.

viewshed—The area on the ground that is visible from a specified location.

waste management—All activities associated with the disposition of waste products after they have been generated, as well as actions to minimize the production of wastes. DOE has defined waste management to include waste storage, treatment, and disposal (but not transportation), and the term is used interchangeably with "waste operations" in DOE planning documents.

water quality—(i) The fitness of water for use; and (ii) the physical, chemical, and biological characteristics of water.

water right—A legal right to the use of water.

weathering—The breakdown of rock through a combination of chemical, physical, geological, and biological processes. The ultimate outcome is the generation of soil.

well—An artificial excavation (pit, borehole, tunnel), generally cylindrical in form and often walled in, sunk (drilled, dug, driven, bored, or jetted) into the ground to such a depth as to penetrate water-yielding rock or soil and to allow the water to flow to or be pumped to the surface or recharged into the subsurface; a water well.

wetland—The U.S. Army Corps of Engineers and the EPA define wetlands as those areas that are inundated or saturated by surface or ground water at a frequency and duration sufficient to support, and that under normal circumstances do support, a prevalence of vegetation typically adapted for life in saturated soil conditions. Wetlands generally include swamps, marshes, bogs, and similar areas.

wind rose—A graphic display of the distribution of the wind direction at a location during a defined period. It is a set of wind statistics that describes the frequency, direction, force, and speed. The characteristics patterns can be presented in either tabular or graphic forms.

APPENDIX G
COMMENTS ON DRAFT NUREG-1748

PAGE INTENTIONALLY BLANK

Background

On October 18, 2001, the U.S. Nuclear Regulatory Commission published a notice in the *Federal Register* (66 FR 52951), to notify the public of the availability of draft NUREG-1748, "Environmental Review Guidance Associated with NMSS Licensing Actions," and to solicit comments. The document was issued for interim use and comment, with comments requested by September 30, 2002. Due to problems with availability, the NRC subsequently published another *Federal Register* notice on May 29, 2002 (67 FR 37461), to extend the comment period to November 30, 2002. Comments were received from one external organization.

Comments Provided by Dominion Generation, Dated September 10, 2002

Comment: NUREG-1437 and NUREG-1555 both provide for review of design-basis accidents and severe accidents from an applicant's ER or FSAR, and review of SAMA analyses from an applicant's ER. Although this NUREG-1748 guidance document states environmental impacts from accidents are addressed, it is not seen in the Contents or directed in the text, specifically. The purpose for mentioning design-basis events above is not clear. It is suggested therefore, that either accident and/or mitigation analyses discussion be included or directed specifically for consistency of applicant expectations with other NRC environmental documents, or additional explanation be provided for disposition of accident analyses.

NRC Staff Response: The purpose for mentioning design basis-events (for 10 CFR 72 licenses) is to indicate a type of accident (i.e., "reasonably foreseeable") whose environmental impacts would be analyzed in an environmental review document.. Stated another way, beyond design basis events (for 10 CFR 72 licenses) and their potential environmental impacts would not be analyzed in the environmental review document as they are typically not considered "reasonably foreseeable." The scope of this document does not provide any additional guidance on accident evaluation or the preparation of safety evaluation reports and is left to existing guidance within NMSS.

Comment: It is not stated that reviewers should identify and evaluate the quality assurance measures taken by the applicant in collecting and analyzing data. This is provided as an instruction to reviewers in NUREG-1555. It is suggested that inclusion of this type of instruction would serve to build public confidence in NRC processes and applicant credibility.

NRC Staff Response: Wording will be added to indicate that licensing and environmental PMs should consider quality assurance measures taken by the applicant in collecting and analyzing data.

Comment: It is suggested that a brief discussion of NUREG-1437 content applicability and usefulness be included in this paragraph. It would serve to further strengthen and clarify the explanation of adopting generic issue findings, particularly as the process applies to spent fuel storage license renewal actions.

NRC Staff Response: The section on "Tiering" will be updated to provide an example of using NUREG-1437.

Comment: It is suggested that NUREG-1437 and Supplement I to Regulatory Guide 4.2 be added to the list of references, to provide reviewers and applicants additional reference guidance regarding license renewal actions.

NRC Staff Response: The reference section is related to documents cited in the text. NUREG-1437 will be added to the references section (see above response). Supplement I to Regulatory Guide 4.2 describes acceptable format and content guidance related to the application for the renewal of a nuclear power plant operating license submitted pursuant to 10 CFR 54, "Requirements for Renewal of Operating Licenses for Nuclear Power Plants." While the general methodology is acceptable (e.g., calculating air impacts in a non-attainment area) the scope is generally not acceptable for NMSS licensing actions.

Comment: For existing licensed facilities, the above-listed resource areas with low-impact (or no-changes) should be deleted from evaluation for license renewal and continued operation.

NRC Staff Response: It is noted that certain issues have been generically addressed in NUREG-1437 and codified in 10 CFR 51, Appendix B. For certain NMSS licensees (e.g., 10 CFR 72) there may not be a need to address these issues. However, this guidance document is written generally to encompass all NMSS licensing actions, hence the statement at the beginning of Chapter 6, *"The Environmental Report: Format and Technical Content,"*

> "It may not be necessary for every resource to receive the same level of detailed review and every action may not require all the information discussed in this chapter."